D1576965

Home-Grown
Fruit

Jane Eastoe

Home-Grown Fruit

*Inspiration and practical advice
for would-be smallholders*

First published in the United Kingdom in 2007 by
Collins & Brown
10 Southcombe Street
London W14 0RA

An imprint of Anova Books Company Ltd

Published in association with The National Trust
(www.nationaltrust.org.uk) and The National Magazine Company
Limited. Country Living (www.countryliving.co.uk) is a trademark of
The National Magazine Company.

Commissioning Editor: Miriam Hyslop
Design Manager: Gemma Wilson
Illustrator: Carmen Carreira-Villar
Designer: Bill Mason
Editor: Jennie Buist Brown
Senior Production Controller: Morna McPherson

ISBN 978-1-84340-416-3

A CIP catalogue for this book is available from the British Library.

10 9 8 7 6 5 4 3 2

Reproduction by Spectrum Colour Ltd, UK
Printed and bound by WS Bookwell, Finland

This book can be ordered direct from the publisher.
Contact the marketing department, but try your bookshop first.
www.anovabooks.com

HOME-GROWN FRUIT

FOREWORD   6

WHY GROW FRUIT?   8

YOUR FRUIT GARDEN   14

BASIC CARE   21

A YEAR IN THE FRUIT GARDEN   33.

FRUIT VARIETIES   49

COMMOM PROBLEMS   81

USEFUL ORGANIZATIONS   90

BIBLIOGRAPHY AND FURTHER READING   92

GLOSSARY   93

INDEX   95

FOREWORD

Welcome to *Home-Grown Fruit* which, along with *Home-Grown Vegetables*, *Beekeeping* and *Henkeeping*, is one of the books that make up this new *Country Living* and National Trust series on becoming more self-sufficient. Nowadays, we are all very concerned about where our food comes from, how it has been produced, how far it has travelled to reach the supermarket shelves and, most importantly perhaps, how the environment may have been damaged in the process. The National Trust helps and encourages farmers to reach and maintain the highest environmental and animal welfare standards, and both the National Trust and *Country Living* champion local, organic food and sustainable practices by British growers. Now, with help from these excellent guides you, too, can begin to produce enough organic and tasty food in your own garden or allotment – honey, fruit, vegetables and eggs – to help feed you and your family throughout the year.

We have all become used to eating fruit out of season but, as Jane Eastoe points out in this excellent guide to growing your own, what we have gained in convenience, we have lost in the taste stakes. Juicy strawberries, picked ripe from your own plot in June are far superior to those hard, tasteless berries from the supermarket – even if you can get them in January. Some things are worth waiting for and, if you grow some fruit for yourself, you will really experience the taste difference. And, says Jane, growing your own fruit is remarkably easy, does not take over the entire garden, is not labour intensive and is the perfect introduction to self-sufficiency for the idle gardener, harassed parent, or time short enthusiast; as fruit trees, bushes and plants are remarkably tolerant and will go on fruiting even when neglected. You and your family will rediscover the pure pleasure of delicious ripe fruit picked straight from the plant and any excess can be turned into delicious jams and chutneys. Full of practical information and beautiful pictures, this excellent guide by Jane Eastoe will inspire you to join the ranks of people up and down the UK who are reaping the benefits of growing their own fruit.

Susy Smith
Editor, *Country Living*

John Stachiewicz
Publisher, The National Trust

WHY GROW FRUIT?

Browse around the supermarket and you will find serried ranks of the most beautiful identikit fruit, all shiny and picture-book perfect, blushing with uniform ripeness. There is a startling, year-round availability of virtually every fruit on the planet. If you want strawberries, peaches, plums or raspberries out of season, then they are available at a price. The cost impacts on more than the purse; the taste, in my experience, is always compromised. Should we be consuming hard, tasteless strawberries in January just so that we can indulge our desire to eat what we want, precisely when we want it?

THE REALITY IS that we do and, as a result, have become accustomed to the bland tastelessness of supermarket fruit. Pears and nectarines are hard as bullets and cold. Why? Because they have been picked before ripening, flown half way around the world and refrigerated to prevent them spoiling. Strawberries, similarly, are cold, overly firm to the touch, taste like an old tissue and transform, infuriatingly, from under ripe to mildewed in a matter of a few hours. Plums no longer dribble juice when you bite them. The fact that some supermarket fruit never reaches a point at which it is delicious to eat should serve as a reminder that there are sound and sensible reasons why fruit should, ideally, only be eaten when in season.

Most commercial growers are, quite understandably, primarily interested in fruit varieties that are disease resistant, highly productive and which have a good shelf life – the question of taste is far less significant. It would be unfair to blame this on growers and supermarkets alone. European Union directives relating to apples, for instance, dictate pre-requisite size, shape, colour and quality for different varieties that growers must adhere to – taste and fragrance are left out of the equation.

Moreover, even when real care has been taken to ensure that the fruit flown in from distant shores still tastes good, then you can guarantee it will be swathed in polystyrene and plastic protective packaging and in cost terms will move beyond pricey to eye-poppingly expensive. Of course it tastes good, but you need to take out a small mortgage to buy it. You must also consider the size of the carbon footprint – all that packaging and aviation fuel is not good for the environment.

I am not recommending that you never buy fruit from the supermarket ever again; life without citrus fruit, bananas and

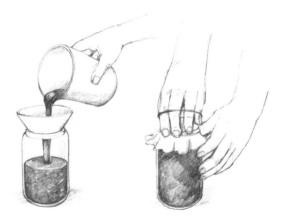

Nothing tastes better than home made jam

the occasional piece of exotica is hard to imagine. However, I am here to encourage you to focus on only eating fruit in season; satsumas in late autumn, forced rhubarb in February and English raspberries from June to October. Believe me, it tastes a whole lot better.

Better still, grow some fruit for yourself so that you can really experience the taste diff e rence. It is remarkably easy, does not take over the entire garden and is not labour intensive.

Everyone knows that a tomato is technically not a vegetable, but a fruit. However, a number of other fruits are commonly regarded as vegetables; namely peppers, avocados, cucumbers, courgettes and peas, which are all actually fruit. The true definition is that a fruit is the sweet ripened ovary of a seed-bearing plant. Whereas a vegetable is a herbaceous plant cultivated for its edible parts, which could be the leaves, the stems, the roots or the tubers. In addition, fruit is commonly sweeter than vegetables because it contains the simple sugar fructose. In this book I am focusing on fruit that is widely recognised as such, rather than abiding by the strict technical definition.

Growing your own fruit is the perfect introduction to self-sufficiency for the idle gardener, harassed parent, or time-short enthusiast; as fruit trees, bushes and plants are remarkably tolerant and will go on fruiting even when seriously neglected. Many plants, once established, will be productive way beyond your lifetime, whilst others will give you anything from five to 15 years before they need replacing.

Whilst the concept of full fruit and vegetable self-sufficiency is appealing, the practical reality is often too much for many of us – it is far easier, less time-consuming and less demanding to be a successful fruit grower. Fruiting plants also have the advantage of sitting comfortably within the average suburban garden. They are not as demanding in space terms as vegetables, nor do they require the same backbreaking annual round of digging, sowing and weeding. Apple, plum, cherry and pear trees will be smothered in fragrant blossom in spring and are far prettier than the overblown ornamental cherries. Damsons and sloes can be incorporated into native hedges. Redcurrants, blackcurrants, raspberries and gooseberries will sit happily in the shrubbery, grape vines will trail over pergolas. Figs, apricots and peaches can hug your walls. Strawberries can be grown in containers if you are short of space.

You and your family will rediscover the pure pleasure of delicious ripe fruit picked straight from the plant. You eat as you go, consuming juicy morsels that taste as intensely strong, sweet or sharp as they should and which smell glorious. It is a curious fact of life that children who refuse to eat fruit from the supermarket cannot resist tasting as they pick their own in the garden. Excess fruit is utilised for jams, chutney or bottling. The family experiments with new desserts to tantalise the taste buds and there is always the freezer which, even in the depths of winter, allows you to enjoy your home-grown gooseberries and relive the glories of summer. Neighbourly relations also benefit from bumper fruit crops as you can happily proffer some of your seasonal glut.

Home brewing can also utilise home-grown fruit; in the early '70s my father had the only listed vineyard in South Wales (though it actually consisted of just the two vines). He attempted brewing everything from elderberry wine to tackling the grape – boosting his home-grown supply with boxes of overripe fruit left by the accommodating local grocer. I had to discard anything rotten or bruised and feed the rest through the ancient food processor. My payment for this labour was to eat as many grapes as my stomach could hold.

Dad even tried making beer (grew his own hops) and gin (an illegal still was erected in the garage for a week and my mother, who had been used as an unsuspecting guinea pig, had to be pushed up the stairs to bed on her hands and knees). The end results were pretty good and at one point he had 250 gallons of wine stored in the garage! Perversely, though, with three apple trees in the garden and a massive annual crop, he never turned his hand to cider.

Fruit, whilst not quite as rich in vitamins as vegetables, is still packed with goodness and blessedly most children will eat it readily, happily consuming their daily recommended quotient of fruit where they have to be force fed each mouthful of spinach. Fruit picked fresh from the garden is at its most nutritious and it will not have been sprayed with chemicals.

Centuries of patient work have built up the repertoire of fruit varieties: strawberries have been around in the UK since the Ice Age and pears grew wild in Britain for centuries before being cultivated. Apples originated in the Middle East reputedly some 4,000 years ago. Apricots and mulberries are from China and plums from Russia. The Romans introduced the grape and varieties of dessert apple to Britain while monks brought new varieties of apple over from France in the Middle Ages, specifically for cider production. The excitement of the renaissance extended even to gardening, broadening the range of fruit cultivation with apricots and peaches being planted against warm walls and espaliers being developed in France. The first orchards in Kent were directed by the fruiterer Richard Harris under orders from Henry VIII. Kent has become known as the garden of England and, to this day, is still packed with orchards. We all collect fruit from other nationalities and delight in growing it ourselves. A neighbour of mine in Kent produced a crop of more than a thousand kiwi fruit this summer.

Fruit growing is also a good way to encourage wildlife into the garden. You can establish a strong ecosystem as everything will benefit from having more birds and pollinating insects in the garden. The birds will consume your crop with every bit as much as pleasure as you, given the chance. However it is worth losing some fruit to them and gaining the benefits they bring in pest control.

Every small contribution we make towards reducing fuel consumption, and the packaging waste that a trip to the supermarket involves, is to the good. Every mini orchard, cropping unusual fruit varieties and promoting a sound ecosystem have a positive effect on the planet, too. What are you waiting for?

YOUR FRUIT GARDEN

How you choose to grow fruit is entirely a matter for you. It is a passion that you can indulge by setting aside a sizable chunk of your garden for major fruit production, or which you can contain by incorporating just a few select fruit trees and bushes into your overall scheme.

IT IS IMPORTANT TO REMEMBER that a large fruit garden is a thing of beauty in its own right. Any well-designed garden will have areas which are stronger in one season than another – a fruit garden should be bursting with blossom in spring and full of fragrant fruit from June to October. It will be an area you are drawn to, to check up on the progress of your budding fruit, to sniff the perfumed air, heady with the scent of ripening fruit or, most importantly, to select what you will be having later for dessert.

A good vegetable garden is laid out with regimental precision with much use of string lines to ensure that nothing grows out of alignment. A fruit garden is softer and more relaxed and, although each variety of fruit has specific spacing requirements, each plant does not have to be a mirror image of its neighbour. It is also a good idea to cultivate flowers as companion plants in the fruit garden; they act as decoys; attracting predators and distracting pests from your crop, repelling or confusing insects with their perfume. Hoverflies, which help control aphids, are attracted to any flat-headed yellow flowers; nasturtiums (*Tropaeolum*) are irresistible to blackfly and woolly aphids. Alliums are beneficial planted close to apples. Fennel (*Foeniculum*) and the California poppy (*Eschscholzia californica*) are very attractive to insects. A fruit garden is ideally situated in a sunny position and, for the absolute perfectionist, is sited on a gentle, south-facing slope so that the soil heats up quickly in the spring. It is important that a working garden is easy to use and it is therefore helpful to have plenty of paths running through it, but this only serves to add to its appeal – children love racing through the paths with tall fruit bushes to hide behind and espaliers and cordons acting as screens. Ensure that paths are no less than one metre wide; you'll want to be able to push a wheelbarrow through comfortably. Raised beds help to contain

Clockwise from above: bush, cordon, espalier, fan, forked cordon, pyramid

soil, keep paths soil free and make it much easier to hoe between your fruit crop.

However, if you are devoted to your flowers and just want to add perhaps two or three fruit trees and some currant bushes, you can still enjoy bumper crops and that special home-grown taste with just a few plants. Try to pick fruit that won't all crop at exactly the same time – perhaps raspberries, a plum and a late apple – so that you have a steady supply of fresh fruit throughout the summer. If you intend to put a few fruit trees or bushes in one limited area, you may wish to plant specimens with similar nutritional requirements; plums, cherries, blackberries, loganberries and blackcurrants need lots of nitrogen, while apples, gooseberries and redcurrants need plenty of potash.

SOURCING VARIETIES

One of the great pleasures of planning a fruit garden is poring over catalogues from different nurseries. Most fruit is best planted between November and March so you can hunt for the specific variety you desire from the comfort of your deckchair. It is advisable to buy your fruit from a specialist nursery. If you buy the first apple tree you come across you could be kicking yourself in a few years time; that rash purchase could lead to years of fruitlessness. Take your time to select what you want and consult the specialists.

It is worth making a rough plan on paper, noting spacing requirements so that you know what you can and can't fit into your garden. It is not worth squashing a few extra plants in as your crop will only suffer for being squeezed. Remember that one of the main reasons for pruning is to open up the plant structure to allow light and air to circulate freely, if you cram plants in you are simply shooting yourself in the foot!

Similarly don't go crazy and start to grow so much of everything that you have in absurd excess just because you have the space. Betty Macdonald, in her book, *The Egg and I*, describes her overly fruitful experiences on a chicken farm in the 1930s: "Towards the end of June Bob and I made several pilgrimages to the abandoned farm and picked five gallons of wild blackberries and the canning season was on. How I dreaded it! Jelly, jam, preserves, canned raspberries, blackberries, loganberries, wild blackberries, wild raspberries, apple sauce, peaches, pears, plums, rhubarb and cherries. By autumn the pantry shelves would groan and creak under Nature's bounty and the bitter thing was that we wouldn't be able to eat one tenth of it. Canning is a mental quirk just like any form of hoarding. First you plant too much of everything in the garden, then you waste hours and hours in the boiling sun cultivating; then you buy a pressure cooker and can too much of everything so that it won't be wasted."

The range of fruit on offer is daunting but, as if that were not enough, you will also have to decide what style of tree suits you best. Standards have a trunk of about 180-210 centimetres high and need about nine metres of space; they are best suited as specimen plants in the lawn or, if you are lucky enough to have the space, in a small orchard.

Half standards are smaller trees, with trunks of around 120–180 centimetres high; this means the fruit is blissfully easy to prune and crop. I have a wonderful old book that recommends them specifically for use in poultry runs, on the basis that they provide the birds with shade but have branches sufficiently high to prevent the hens from getting at the fruit. Don't let this recommendation hold you back if you don't keep chickens. So much better to have a half standard apple or plum than a cute ornamental pear which, though pretty, merely produces hundreds of minuscule, rock hard fruit of neither use nor ornament.

Apple and pear bushes have 70 centimetre high stems and plums and cherries have one metre high stems. Such trees are generally grafted onto dwarfing stock to keep them small. New varieties of peach have been grown onto dwarf rootstock enabling them to grow well in containers, making fresh peaches a more realistic concept for those of us without glass houses. Some soft fruit are grown as bushes and, in the case of redcurrants and gooseberries, the stem should be 10 centimetres high.

A cordon is a tree with a single stem. Side growths are pruned back hard and thus short fruiting spurs are encouraged right up the stem. A cordon takes up very little room and has the added advantage of allowing a number of different fruit varieties to be grown in a small garden.

Cordons are ideal if your space is limited

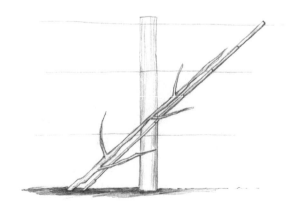

It is usually planted upright and then trained to grow at an angle of 45 degrees. It is possible to train a cordon to grow parallel to the ground (known either as a double horizontal cordon or a one storey espalier) or to have a double stem, triple stem, or a forked cordon (which looks like a vertical espalier).

An espalier has five or six tiers of horizontal branches trained to grow at a distance of 30 centimetres apart and at right angles to the main stem. This is well suited for growing apples and pears. Fan-shaped trees have a main stem of about 45 centimetres high and the branches grow out from there. This form is well suited to apricots, cherries, figs, nectarines and plums.

It is better to plant a young tree than an older one, although it can be advantageous to have one that has been pruned by an expert for the first two years of its life. A young tree gets over transplanting faster than an older one and will quickly settle into its new position. When purchasing bush

*An espalier can rival any
flowering shrub in looks*

*Fan train tender fruit
against a warm wall*

trees, they should be one to two years old, half standards
should be three years old and standards three to four years
old. Cordons should be two years old and espaliers three
to four years old. Currants and gooseberries are generally
bought as two-year-old bushes or cordons. Blackcurrants can
be one or two years old while one-year-old rooted tips are
right for blackberries and loganberries. Raspberries should
be young canes.

BASIC CARE

Successful fruit growing is initially dependent on two factors being correct: situation and soil. If you plant your chosen fruit in the right situation, generally sunny and open, and your soil has been dug over and fed and the drainage is right, your plant will thrive. Please remember that gardening need neither be difficult nor time consuming, what you put into it will affect what you get back, but being short of time needn't mean that you can't grow fruit – it just means that your fruit might not be as big, shiny or competition worthy as that of your neighbour. What counts is that you, your friends and family can enjoy the many benefits that growing your own can bring. I would urge you to give fruit growing your best shot and if you don't manage to do everything according to the book, it doesn't matter. Next year you may do better; plants are very forgiving, they want to flower and fruit and will do their best to provide you with a generous crop.

THE SECOND MAJOR ADVANTAGE of growing your own fruit is that you can control exactly how it is grown. With the exception of organic produce, all the other fruit that you purchase will have been sprayed with pesticides and fungicides and covered in chemicals. Your own fruit can be grown according to organic principles and can be just as good as nature intended. You may have the odd bug hole, but what you are eating is not only delicious but positively good for you, too. You grow it, you pick it and you eat it – what could be better or more natural? I'm afraid that food only comes bug free at a cost!

Good soil condition is the starting point for any organic gardener. Healthy plant growth is dependent on the quality of the soil and plants need to be fed, just as we do. A good rich soil is full of humus; decomposing organic matter bursting with insects and micro-organisms. Your soil may be lacking in nutrients, but it is easy to improve the quality, or indeed just to give your soil a healthy boost in specific areas.

Composting is the most economical way to improve soil condition. It costs nothing, if you don't count the price of buying or building a compost bin – a worthwhile expense because it makes the whole process much faster and is more efficient than an out-of-control compost heap. Composting also makes sound ecological sense – none of us should be discarding green waste that we can recycle.

Sensible and realistic plant selection will also make the difference between success and failure – why struggle to grow lemons, grapes or figs if you live in a cool, damp area? Select plants that will thrive. Consider raspberries, blackcurrants and gooseberries, which will be far easier to grow. Organic gardening works with nature, not against it. It will help you build up a sound eco-system in your plot and, by encouraging wildlife, you are enabling nature to control many of the pests that commercial growers use chemicals to eradicate. It is important to remember that those infuriating caterpillars that attack your greenery will turn into pollinating butterflies and that the sprays you use to dispose of aphids will also kill ladybirds as well – and your next aphid infestation will be infinitely worse without ladybirds to assist. Where insects come, birds, bats and hedgehogs will follow, all of whom will help to improve the condition of your soil. For more information see the chapter on Common Problems.

CULTIVATION

The starting point of soil improvement prior to planting is the removal of weeds. Weeds not only compete with cultivated plants, they also harbour pests and diseases. Thorough digging is the most effective way to eradicate them; it helps to remove the roots of pernicious perennial weeds such as bindweed, nettles, dandelion, couch grass and ground elder. It really is worth taking some time to do this properly, the tiniest piece of root left in the soil will spawn a new weed. Moreover, you will only have to dig the soil effectively once; after that a little hoeing should ensure your ground is weed free.

SOIL

Soil is graded according to its clay, silt and sand content. The proportion and size of these mineral particles affect the behaviour of the soil.

LOAM SOILS: Generally regarded as the most desirable with the perfect combination of mineral particle sizes and around 10–25% clay – a mix that offers high fertility combined with good drainage and water retention.

CLAY SOILS: Heavy and slow draining often with a high nutrient content. It is easy to damage the soil structure by compacting it and the soil is slow to warm up in the spring.

SANDY SOILS: Light, free draining and blessedly easy to work. They also warm up quickly in the spring. They are not very fertile although this problem can easily be rectified by adding organic matter and feeding and irrigating frequently.

SILT SOILS: More fertile and moisture-retentive than sandy soils and they also warm up quickly in the spring. The downside is that the soil structure is easily damaged by compaction.

CHALK SOILS: warm up quickly in the spring, they are moderately fertile and free draining, however they tend to be stony and rather shallow to work.

Most soils are about 30 centimetres in depth, though this is not always the case: in parts of Kent, the soil is only 10 centimetres deep.
The pH balance of a soil is a measure of its acidity or alkalinity on a scale of 1–14. A reading below 7 indicates an acid soil while a pH reading over 7 indicates an alkaline soil. A neutral soil has a pH balance of between 6.5 and 7.3. If your soil is very acid you can raise its pH balance by adding lime to sweeten it and if it is very alkaline you lower the pH balance by adding sulphur or by adding nitrogen fertilisers. Simple pH testing kits are readily available from garden centres and are easy to use.

FOOD SUPPLEMENTS

Hungry fruit will make extra demands on your lovingly prepared soil. Although you must not dig the area over on an annual basis, you will still need to provide your fruit with occasional food supplements – though it very much depends on the variety as some fruit put on too much growth if fed over enthusiastically.

Organic fertilisers are rated according to their nitrogen (N), phosphorus (P) and potassium (K) content. Nitrogen stimulates growth and leaves, phosphorus benefits the roots and potassium boosts disease resistance and fruiting. Choose carefully to suit your plant's particular requirements. Apart from compost and farmyard manure (FYM) which can be applied as a top dressing, organic fertilisers come in the form of pelleted chicken manure and liquid feeds made from comfrey (*Symphytum officinale*) or stinging nettles (*Urtica dioica*). Wood ash from log-burning stoves is a source of potash and can be put directly onto the soil to help the plants immunity against diseases and to encourage ripening – gooseberries and apples are particularly partial to this.

FYM

Good farmyard manure is known as 'sweet' and it contains well-rotted straw and is packed with nitrogen, potash and phosphates. This is the natural way to improve the pH balance of your soil and to give your plants the correct nutrition.

COMPOST

Garden clippings and green household waste; egg shells, tea bags, coffee grounds, discarded vegetable leaves, beans, lentils, rotting fruit, old clothes – so long as they are made of natural fibres and chopped up – and newspaper are the principal ingredients of a good compost mix which will, in time, rot down into a nutritious soil-like material. To keep the composting process going add an activator such as chicken manure, urine (I kid you not, get out there men) or seaweed. Seaweed has the added benefit of being a good source of lime which is essential for the growth of stone fruit. Even 'sweet' farmyard manure benefits from some time in the compost heap. Wood will take a very long time to rot down in compost – it helps if it is shredded first. Big pieces of wood, anything smothered in thorns, or diseased leaves should not be composted – burn them on a bonfire instead and use the ash as a food supplement added direct to your fruit (see page 25), or put the ash into the compost. Never put meat or dairy produce onto your heap as this will attract vermin and avoid putting perennial weeds on too; they may well survive the composting process and take the opportunity to colonise a whole new section of the garden.

A good compost bin is worth its weight in gold, if it is big enough it will generate sufficient heat to speed up the composting process – a good bin should be about one metre square and well ventilated with good air circulation and drainage.

Composting works when mixes of different materials are combined with air. As the mix starts to break down it begins to heat up – a good compost heap will steam when it is turned. Regular turning helps to ventilate the heap and improves the cooking – think of a log-burning stove which roars into life when the vents are opened. A well-insulated compost bin can exceed temperatures of 60 degrees Centigrade (140 degrees Fahrenheit) so make sure your bin is lined with old carpets or newspapers and make sure it is covered at the top.

When the compost is well cooked, leave it to mature for around six months before using. This is why it is helpful to have two or three compost bins on the go at the same time. The most common reason for failure is an insufficient mix of materials, or the mixture being either too wet or too dry. The remedy is to remake your heap and add more moisture or dry material as required.

It is also possible to cultivate your own green composts: sow seeds, watch them grow and then, after a few months, hoe or dig it all back into the soil; the poached egg plant (*Limnanthes douglassii*) and buckwheat (*Fagopyrum esculentum*) can be used to good effect. These are best used when the patch is left fallow for a few months. If you prepare your fruit patch in late spring for planting in October, why not sow a crop of poached egg plant and hoe it into the soil in September, along with FYM, a month before planting? You are all prepared to plant, but won't be looking at an expanse of bare earth all summer.

MULCHING

A mulch is a material applied to the top of the soil which conditions it and smothers weeds, acts as a barrier to weed seeds, or inhibits moisture loss. Strawberries, for instance, need annual conditioning mulches, but often additionally have sheets of black plastic on either side of plant rows to discourage weed seeds. Compost, manure or even a few centimetres of grass clippings will all make good, conditioning surface mulches or top dressings; they won't control weeds but will improve the quality of the soil. Porous plastic membranes keep surfaces weed free after clearing but allow water to drain through, while sheets of black polythene or old carpet will smother and suppress weeds very effectively – but this method is no quick fix, it will take a year or more to be effective. Old newspapers can also be used; worms will pull them into the ground little by little!

DRAINAGE

Many fruit trees and bushes prefer well-drained soil and so it is important to try to ensure that your soil is reasonably well drained. Heavy clay soils in particular are inclined to become waterlogged; the water forces air out of the soil and it this that severely damages plants. Incorporating some sand and organic matter will help improve soil structure. Alternatively, if your soil is very problematic, you may wish to build a raised bed. This method will allow you to create your own perfectly draining patch, all topped with a wonderful loam.

IRRIGATION

The best way to irrigate plants is via the soil as many fruits are susceptible to water damage by rain or by sprinkler. Furthermore you are better leaving plants without water than giving them a quick once over: save watering for when you can devote a little time to do it properly and soak the soil – hosepipe bans permitting. Water butts, as long as they are used and emptied frequently so that water can't stagnate, are helpful, as is utilising the water from your bath. Do not siphon off water from the washing machine or dishwasher as this will contain too many chemicals.

SITUATION

The sun rises in the east and sets in the west, therefore south and southwest-facing plots and walls get the most hours of sunlight. North-facing gardens and walls are the coldest and shadiest. The majority of fruit does best in an open, sunny site and therefore it is very important that you work out the best place for your tree, bush or indeed fruit garden.

If your garden has any low lying spots you might have problems with frost pockets. Warm air rises, but cold air will flow to the lowest point, it moves like water and once it has hit the low point it fills it up with cold air. When frosty air reaches the branches of trees and bushes it will kill the blossom.

It is simple enough to erect a post and wire support for espaliers and cordons. Either use a ground-awl to help you push your stakes deep into the ground, or use concrete foundations for extra strength. The post should have a finished height of 1.8 metres. Put posts at 3 metre intervals. Use straining bolts at each end and run three or four strand wires at 30 centimetre intervals with the base wire 60 centimetres from the ground. If you are wiring a wall to take an espalier or a fan-trained tree, you will require a system of vine eyes, wire and strainers to pull the wire tight. Vine eyes should generally stick out about 6–7.5 centimetres from the wall in order to allow a current of air between the wall and the tree. The first line of wire should be 60 centimetres from the ground and f u rther rows should be spaced at 30 centimetre intervals.

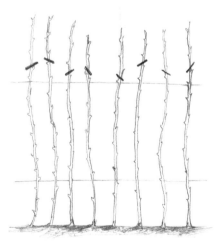

Nip the top off raspberry canes every February

Raise new strawberry plants from runners

Most trees should be staked until they have established a support system; until then, they are liable to rock on their roots, which can cause damage. Bush trees should be given a diagonal stake while standards and half standards should have a vertical stake. It is important to use proper tree ties and spacers when staking to ensure that the tree is not damaged by an unsuitable tie and also to prevent it from rubbing against its stake in the wind. Always put the stake in front of the tree in the direction of the prevailing wind, so that the stake holds the tree in position, rather than vice versa. Remember to check that your ties are not too tight every few months; your tree is growing and the tie should not cut into it. If you keep pet rabbits, or if you live in the country and have wild ones in the garden, protect your tree at the base with chicken wire to prevent them gnawing at the bark.

PLANTING OUT

Most trees are best planted out when in their dormant period, between November and March; the earlier the better is true in this instance, as it gives the trees a chance to settle in before winter gets going.

If your trees are bought bare rooted, it is imperative that they are planted out as soon as possible. If you do not have time to attend to this immediately, you can always heel the tree in. This is a temporary planting: you dig a hole and rest the t ree at an angle in the hole so that its trunk is resting on the e a rth. Then dig out more soil from the far side to cover the roots. This ensures that the roots are not damaged by exposure to the air and that the tree is well supported and not stressed.

However hopefully you prepared the site a month or two earlier, it is always worth digging it over and adding plenty of sweet farmyard manure. Do not dig the hole in advance, as this can dry out or become water logged. Dig a hole to a depth of approximately 25 centimetres and a diameter of

Plant promptly or heel in as a temporary measure

80 centimetres. Do not make the hole too deep; the graft of the tree must be above the level of the soil – keep checking the level is correct. Spread the roots of the tree out so that they look like the sun's rays (trim off any damaged ones) and start to cover with soil, treading in well as you go. Keep filling and treading to ensure no air pockets are left.

A YEAR IN THE FRUIT GARDEN

Life in the fruit garden is not the frenzied round of activity that is seen in the vegetable patch. Forget that relentless to do list: crop rotation, digging, feeding, sowing, pricking out, planting, pest control and hoeing. Fruit growing is a far more leisurely affair. Once your fruiting plants and trees are planted out, there are just two annual activities to occupy you. Most trees and plants, although not all, benefit from an annual feed and most appreciate some kind of light annual prune. You will reap the rewards if you keep pests away from your precious crop. See the chapter on Common Problems for more information. Of course, both vegetables and fruit require cropping, storing and preserving – and the cakes, desserts, jams and sauces you can make with fruit are thrilling and delicious.

FEEDING

If plants are to make healthy growth, they must absorb essential chemical elements: oxygen, carbon and hydrogen are absorbed from the atmosphere and the soil, and other nutrients are taken up by hairs growing towards the tips of the roots. Plants depend principally on nitrogen, phosphorus and potassium (see the chapter on Basic Care for more information) and this is available from garden compost, farmyard or organic manure and concentrated feeds such as bone meal and dried blood.

This essential food should not be dug in annually, as this would disturb the root system. Instead it is applied in spring as a mulch around the plant or, in the case of a tree, spread under the full extent of the leaf canopy so that all the roots will benefit. Some fruiting plants, such as the fig, should not be fed as they will put on too much leafy growth; look in a specialist fruit book to see if an annual feed is required.

PRUNING

If anything stops people growing fruit, it is the idea that pruning is a highly complex and technical task without which they will get no fruit. Pruning is, at its best, a skilful craft, but it is easy enough to learn and, to be honest, most fruit trees are very forgiving and will go on fruiting for many years left to their own devices. If you prune something incorrectly – cut off all its fruiting buds, for instance – it won't keel over, it just won't crop very well that year and you will know better the following year! If in doubt, the best advice I can give you is to prune gently.

Pruning fruit trees is done for three main reasons: firstly to encourage the growth of fruiting wood; secondly to open up

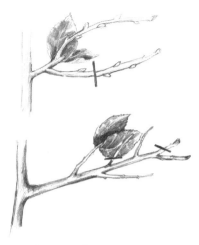

Tip pruning and spur pruning an apple tree

the framework of a fruit tree or bush so that branches do not cross each other, thus allowing air and sun to penetrate and ripen the fruit and discourage disease; and thirdly to control height to make it easier to crop. The basic principles are that when you prune a tree hard you are promoting the production of leaves and branches and when you prune it lightly you are encouraging the formation of flowers and fruit.

A tree will focus its energy on producing roots, branches and leaves; it wants to grow before it will fruit. Upward-growing branches are more vigorous and take longer to fruit than horizontal branches.

If you buy a tree of around four years old that has been trained by a nursery, you will probably not need to touch it for several years. The older varieties of fruit tree that have been around for centuries need less pruning, so if you want to minimise your workload you may wish to purchase one of these instead of a more modern variety. Many commercial growers are looking into returning to older varieties which require fewer man hours for pruning and which are more disease resistant, thus cutting back on the need to spray.

In purchasing an old-fashioned variety, not only are you making life easy for yourself but you are also keeping a fruit-growing heritage alive and maintaining the balance and variety of the fruit on offer. After all, why grow Cox's Orange Pippins

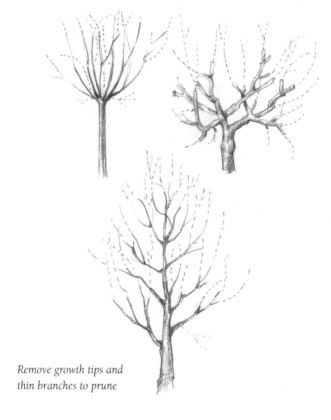

*Remove growth tips and
thin branches to prune*

when you can buy them by the ton in every fruit shop across the country? Instead, look for something different like a sweet and crisp Worcester Pearmain.

One of the first tricks for the fruit-growing novice is to get to know your tree in its first few years in the garden – you will soon learn to recognise the age of different wood and to recognise the leaf and flower buds and the leaders and laterals. It will then be much easier for you to understand the tree's pruning requirements. You will quickly be able to identify the growth buds, which lie flat against the stem, are smaller and more pointed than fruit buds which are rounder and plumper. Check to make sure that you know how your fruit tree likes to be pruned as different fruit have different requirements; plums, for instance, should not be pruned in winter as it makes them vulnerable to disease.

HOW TO PRUNE

It is very important that you have the right tools for this job, as you will damage your tree and frustrate yourself if you cannot cut cleanly and efficiently. You do not want any jagged cuts as this will encourage disease.

The term a hard prune suggests that existing leaders are cut back by three-quarters of their length. Light pruning indicates that leaders are cut back by about one-quarter.

When you want to promote new growth, make a cut above a growth bud. Cuts should be angled (at approximately 30 degrees), do not cut horizontally. Do not cut too close to the bud or you may damage it. Slant your cut away from the bud so that any water will be carried from it, not directed into it. When removing shoots or buds make a cut flush with the bark, do not leave stubbly ends. It is best to use a pruning knife for this work.

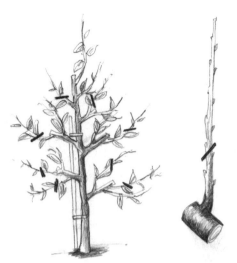

Angle pruning cuts above a growth bud

If you are taking out a large branch, do not cut it all in one go; the branch is likely to tear, which will make the tree more susceptible to disease. Your aim is to have a smooth finish and you will need to cut the branch in two sections to achieve this. First remove the branch at around 30 centimetres from the trunk; start by cutting from underneath and then finish off sawing from above. This avoids weight from the full length of the branch making the cut tear. Cut off the remaining length, so that it will be flush with its parent branch or trunk, again cutting first from below.

WINTER PRUNING

The main purpose of winter pruning is to cut back the growth of the most vigorous shoots and to encourage the growth of side shoots. If this process is carried out when the tree is dormant in winter, there will be no dominant shoots left to grow away in spring and the overall growth and fruiting of the tree will have been encouraged. It is easiest to identify the dominant shoots at the end of the winter and this is also a good time to spot dead or diseased wood and crossing branches that need to be removed.

Don't prune too early, or too late in the season and never when the tree is in flower or when it is losing its leaves, for it will be under more strain at these times than any other. Trees that have been fruiting for 20 or 30 years may benefit from a very hard prune in winter to encourage the growth of strong new leaders and laterals. Do not do the whole tree in one year, however – rather do it little by little over three years.

SUMMER PRUNING AND THINNING

Some fruit trees, such as apricots, cherries and some varieties of apple and pear, require pruning in summer to focus energy on fruiting instead of further growth. Four main tasks are highlighted; thinning involves the removal of 5–10 centimetre shoots, or those that are growing in peculiar directions. Top pruning involves removing a chunk of the tallest branch. Suckers that develop around the base of the tree should be removed as these sap its energy and can cause instability in the root system. Finally, and this is the hardest rule to follow, your precious crop should be thinned to ensure that each fruit has sufficient space to grow – you will ultimately have a better crop for doing this.

PRUNING HARD FRUIT

In winter remove diseased wood, crossing branches and anything that is rubbing, to let in light and air. As the tree becomes more mature, cut leaders back by a third and shorten spur shoots. Always leave fruit buds and one or two other buds on the lateral. Better to under prune than to over prune! Apples, sweet cherries and pears carry fruit on wood that is two years old or more. The aim in pruning is to maintain the ratio between old and new wood. Monty Don, in his book *The Prickotty Bush*, describes some advice he was given by an old boy on pruning his apple orchard: "Cut off all the whippy new growth and sure as you lets in a bit o' light and a bit o' air and you won't go far wrong. Imagine there's a pigeon flying right through the centre of that tree. 'E shouldn't touch no part of it with his wings as 'e goes by."

PRUNING STONE FRUIT

Prune stone fruit in June, July and August when the wounds will not allow the spores of silver leaf disease to enter. Peaches and sour cherries bear fruit on one-year-old wood. Plums and damsons produce fruit on both new and old wood, the pruning aim is to produce plenty of new wood.

PRUNING FRUIT BUSHES

These need to be pruned in a variety of ways; you will need to look at instructions for each particular plant. Redcurrants and gooseberries fruit on wood from one year old and onwards, so the aim in pruning is to produce plenty of new wood.

PRUNING CANE FRUIT

Blackberries, raspberries and loganberries fruit principally on one-year-old wood. Pruning consists of cutting down summer fruiting canes to ground level in the autumn, leaving the new summer's growth to supply fruit the following year.

PRUNING GRAPE VINES

Grapes fruit on the current year's wood. Vines should be restricted to one main stem. Lateral branches are allowed to grow at intervals with other buds being rubbed out. The lateral branches are kept free of side stems except for short growths that carry the flowers and fruit. As soon as the foliage has dropped off in autumn cut shoots back to within two buds of the old wood. As the shoots start to grow in the spring, allow only one of the two to develop. Each shoot is pinched out when three or four leaves develop beyond the flower bunch. If flowers do not develop, the shoot is pinched out when it has eight leaves. Thin the fruits as they develop.

WALL FRUIT

Cordons and espaliers need to have growth checked and directed and fruiting encouraged. This is undertaken in the summer. Laterals of 15 centimetres long that are not required to extend growth are pruned back to within 5 millimetre of their base. Some laterals usually reach this point by mid-June, but this task must be repeated at three-weekly intervals until the end of September. Laterals should not be cut until they have achieved the required length. In October any laterals that reach 15 centimetres can be left on until the following June. The leaders, however, are not pruned back in the summer, they are left until May when they can be cut by a half if they are 30 centimetres long, or by three-quarters of their length if they are 20 centimetres long. This encourages buds to break out at the base of the leaders.

There are many creatures in the garden that will want their share of nature's bounty and who can blame them? Some, like birds, are easy enough to keep out with nets, although this can become a major problem if you have a large cherry or mulberry tree. Old-fashioned methods still work well. Tie pieces of tin foil, CDs, or lightweight plastic bags onto string and hang from poles or tree branches to frighten off birds. Cats are always effective!

Keep an eye open for caterpillars and, if you find some, t ry to get as many off as you can – shaking the plant can be surprisingly effective. If you keep hens, feed them your haul, as they will regard it as a real treat. If slugs are a problem, ring plants with a rough material such as egg shells, wood ash or sand. Alternatively, lay traps: orange halves, or saucers of beer, bran or milk all work well, although then you will have the grisly task of collecting your unwelcome guests and removing them from the vicinity!

Most forms of predator control are designed for use in enclosed environments such as glasshouses. However, one can be used out of doors. Phasmarhabditis is a parasite that lives in the soil and kills slugs. Used early in the season, it can be very effective in minimising slug damage. For more information on pests, see the chapter on Common Problems.

Preserving

Fruit preserving is one of my favourite pastimes. I may sigh heavily at the prospect when the kitchen is piled high with baskets of fresh fruit, but once I get going there is something indefinably wonderful about making jam and jelly (a clear

sieved jam with no fruit pulp or seeds) and chutney.
Everything smells so wonderful as it cooks and that first taste
of a plum or strawberry jam is like heaven on earth. The
delicious sense of smug satisfaction when all is done and your
shelves are groaning surely takes some beating. I drive my
family wild for days waxing lyrical about the latest batch, at a
time when the garden is awash with fruit that can be eaten
fresh. They do appreciate my efforts in the dark days of
winter, though, when the taste of bramble jelly or gooseberry
and elderflower jam reminds them of long sunny days.

Bottled fruit give enormous pleasure and satisfaction, they
look so wonderful stacked on the shelves and opening one
always feels like a treat. A few jars are usually enough to keep
the family happy. There are no limits as to what you can cook
when you have the produce, including fruit cheeses, fruit
butter, ketchup and sauces.

Freezing is much simpler and all fruit, bar strawberries,
can be successfully frozen. The trick with soft fruit is to lay
them all out individually on a tray and, when they are fully
frozen, put them in a bag together, then you can serve them
fresh or cook them later. Alternatively, you cook the fruit and
freeze it so that on days when you are hard pressed, a dessert
is already prepared. I stew pounds and pounds of freshly
picked cooking apples in September, often with some dessert
varieties thrown in for good measure, defrost them and add a
handful of frozen blackberries, gooseberries or raspberries to
the apples to liven up a pie or crumble.

JAM AND JELLY MAKING

Making jam is really incredibly simple and, although it is
delicious on hot toast or muffins, it can also be used to fill

sponges and scones or be dribbled over pancakes and heated to pour over ice-cream. It is so delicious and so entirely different from shop-bought jam that is hard not to simply eat it by the spoonful straight out of the jar!

Select fruit that is just ripe or fully ripe; do not use overripe fruit in jam making as it will have lost much of its pectin content. Wash the fruit and remove stalks, cores and stones – though with damsons you cook with the stones and sieve them out with a slotted spoon as the jam cooks and before potting (an interminable process). Add a little water if required; fruit such as strawberries, raspberries, blackberries, rhubarb and red currants do not need any water added to the mix but less juicy fruits such as plums, apples and apricots need to have half their weight added in water. Simmer gently in a large pan until the fruit is reduced to a pulp.

It is at this point that the sticky question of pectin comes in to the equation. Pectin is the substance that will help your fruit

Make fruit jelly by straining pips

to solidify. Some fruit are naturally high in pectin, such as cooking apples, lemon juice and redcurrants and these fruit are often combined with others to produce a set jam. Cherries, blackberries, raspberries and strawberries have a low pectin content and are unlikely to set on their own. Commercial pectin is available from the shops, but I always prefer to use a mix of fruit to achieve the desired effect. What's more, I never mind terribly if my jam is a bit sloppy – it will still taste glorious. Lemon is generally added as a matter of course to fruit that lose their acid content during cooking: blackberries, peaches, pears, strawberries, raspberries, dessert apples and sweet cherries.

It is important to use the quantity of sugar specified in a recipe, as the amount varies according to the amount of pectin in the fruit. Fruit high in pectin needs to have one-and-a-half times its weight in sugar added e.g. two kilos of fruit to three kilos of sugar. However, if the fruit has a lower pectin content, add the same weight of sugar as you have fruit.

Before you add sugar to the fruit it must be gently heated in the oven to avoid lowering the temperature of the fruit – I always do this before I begin. Add the warmed sugar and cook on a gentle heat, stirring continuously until the sugar has all disappeared and no granules can be seen – this takes longer than you would think. When the sugar has dissolved, turn the heat up and bring to the boil until setting point has been reached – this can take anything from five to 20 minutes. Stir occasionally to stop the jam burning.

To test if a jam is set, put a few saucers in the refrigerator. Drop a spoon of jam onto a chilled saucer and leave to cool – if it wrinkles on the surface when you push it with your finger, it is set.

Your clean jam jars will have been warming in the oven – whip them out and pour in the jam, wiping up any spills on the glass as you go. Put a wax paper circle onto the surface of the jam immediately and put a cellophane circle over the neck of the jar and fix in place with a rubber band. This must be done while the jam is hot to make an air-tight seal.

Jelly is made in more or less exactly the same way, the main differences being that you need to use more fruit proportionately and at the end of the process the mix is sieved to remove pips, pulp, hair, maggots and stalks etc. Fruits suited to jelly making include blackberries, blackcurrants, redcurrants, crab apples, gooseberries and quince, and cooking apples are often added to help the set. When the jelly is made, tie a square of muslin across the top of a sieve, put a large bowl underneath and pour the mix into the sieve, the glorious juices will strain through. Put this into a jug and quickly decant into your warmed jars, seal as for jam.

CHUTNEY

If you have more of a savoury palette, or just an awful lot of fruit, you can make fruit chutney to eat with cheese, cold meats, pies and curries. Chutney is cooked very slowly and needs to mature in the jar – good chutney will last a couple of years. Classic examples include apple and tomato, gooseberry, damson and plum. Recipes usually include white or malt vinegar, an onion and a range of spices and sugars.

Begin by washing and chopping fruit and vegetables very small – particularly onions as they should not dominate. Pop screw top jars into the oven on a low heat to warm. Fruit and vegetables are cooked very gently together in a little vinegar until soft. Then add spices and sugars – light chutneys use

white sugar and dark chutneys brown sugar – and cook very
gently until the sugar is dissolved. Then simmer, uncovered,
for one to two hours until the chutney thickens. The chutney
should be potted in a screw top jar as soon as it is cooked.
Plastic is best, as vinegar corrodes metal. Put greaseproof
rounds on the surface of the chutney and then cover with a
clean cloth that has been treated with melted paraffin wax.

BOTTLED FRUIT

Fruit for bottling must be ripe, fresh and unmarked, overripe
fruit is unsuitable – we are looking for perfect specimens here.
Warm wide-necked jars (with screw or clip lids) in the oven
on a very gentle heat. Make the syrup according to the recipe
for each particular fruit, generally 250 g of sugar to 600 ml
water. Melt the sugar in half the water and when it is
completely dissolved, add the rest. Some recipes use honey
and spices or alcohol (the latter added after the syrup has
been cooked) to give flavour. Pack the jar with fruit carefully
and then pour over the syrup, try to get rid of any air bubbles
by moving the fruit around with a spoon; then put the lids
on loosely.

Sterilise the bottled fruit in a slow water bath. Loosen screw
top lids by one turn to allow steam to escape; clip lids are tailor-
made to do this when closed. Take a large pan, put a thick layer
of newspaper on the bottom and wrap each bottle with
newspaper before putting it in the pan. Fill the pan with cold
water until the bottles are covered and then put on a gentle
heat. It takes approximately an hour to reach boiling point and
then a further half an hour for the fruit to reach its required
temperature. Different fruits need to be cooked at different
temperatures for very specific times, and you must have a

cooking thermometer to check accurately. As soon as the fruit has remained at its required temperature for the decreed time, remove it from the pan, put on a wooden chopping board and tighten screw lids. Leave to cool overnight. The next morning, unscrew the lid and lift it to see if an air-tight seal has been formed: if it has, the whole jar will lift; if not, the lid will come away and the fruit must be eaten very quickly.

DRYING FRUIT

This is one of the oldest methods of fruit preserving. Fruit should be ripe and in good condition. Take a wire cake stand and place on an oven tray. Cover with a muslin cloth and put the fruit on top. Put into a cool oven, around 50–65 degrees Centigrade (120–150 degrees Fahrenheit) or gas mark ¼. Leave for three to six hours depending on the fruit. Remove from the oven. Leave to cool for at least 12 hours, but cover with another muslin cloth first as light will discolour the fruit. The fruit is fully dried if no juice comes out when squeezed. Apples can be peeled, cored sliced, and put into the oven with a stick through their middle. When cooked, leave them to cool for 12 hours before packing away.

Dry apple rings in the oven

Strawberries

Cherries

Apricots

Quince

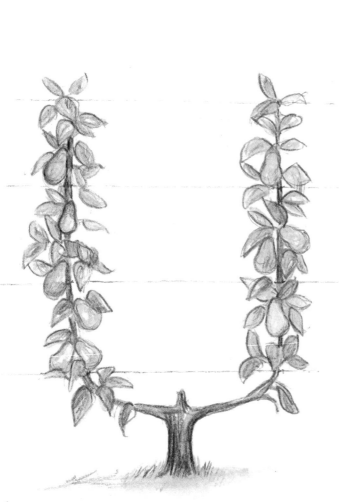

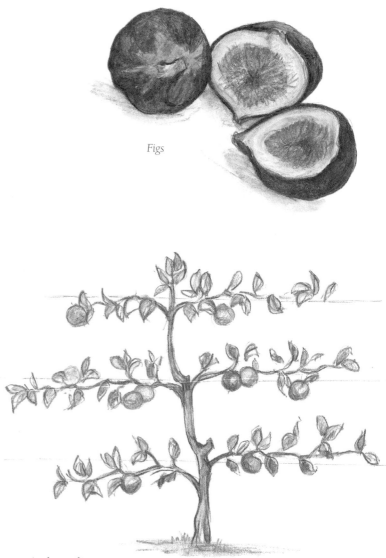

Figs

Pear cordon

Apple espalier

Blackcurrants

Plums

Raspberries

Gooseberries

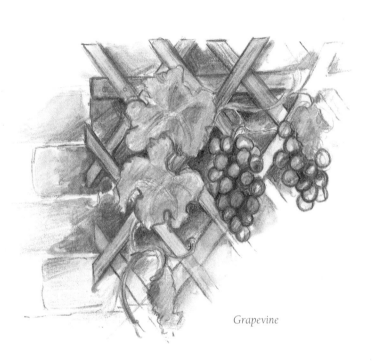

Grapevine

Mulberries

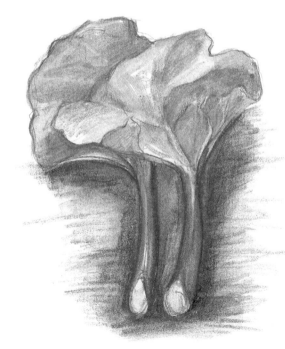

Rhubarb

FRUIT VARIETIES

When it comes to choosing what fruit to grow in your given space, it is worth spending a little time poring over books or visiting specialist nurseries to take advice on which fruit and what varieties will perform well for you. There are, for instance, more than 2,000 varieties of apple alone, so you may decide it is simplest to choose a variety that you know you like to eat. However, some varieties of fruit require cross-pollination – and if yours is the only tree of that variety in the neighbourhood, you will have a fruitless experience! Start by making a wish list of fruit you either find it difficult to obtain, expensive to buy, or which you eat the most of, then see if your space and location can accommodate it.

I T IS ALSO WORTH PAYING ATTENTION to the age of the variety – if it has been around for a long time then you can be assured that it has reasonable disease resistance and will be a good fruiter. In this instance new is not necessarily best, though a new variety with increased disease resistance may make a difference to the size of your crop.

Fruits grow in various forms: on trees as with apples, plums, cherries, figs and peaches; soft fruit grown on canes and bushes: redcurrants, gooseberries, tayberries and wineberries. Trees that fruit will be productive for many years – even old neglected trees can return to fruition after a good prune and a feed. They can be free standing as embodied by the classic lawn tree which can be kept small with regular pruning; trained to grow along wires to provide garden screens; or trained to grow against walls in a variety of forms, the latter particularly useful in small town gardens where every centimetre of space counts.

Soft fruit is a term applied to soft-skinned and juicy fruit which is generally no taller than 1.5 metres, though of course there are always exceptions – grapes will grow very tall given the opportunity. Some plants, such as strawberries, need to be replaced every four years, whilst others, such as raspberries, will last up to 15 years.

Most locations can sustain fruit production, which might sound an ambitious term if you are only thinking of an apple tree and a few strawberry plants! The only environment that could experience difficulties is a garden heavily shaded by large trees; these will take all the nutrition and deny the plants the sunlight and breezes that make them thrive. If you have a problem with an existing fruit tree in your garden that has discouraged you from growing more, then please think again. It is quite likely to be a nutritional problem, see the chapter on A Year in the Fruit Garden for more information. Different fruits are suited to a range of sites, so don't assume that everything requires sun. The Morello cherry, for instance, does best against a north wall. Even if you are limited to a balcony, you can still grow fruit: dwarf root stock ensures that all sorts of goodies, even apple trees and peaches, can be successfully grown in containers.

QUINCE (*Cydonia oblonga*)

DESCRIPTION: This should not be confused with the popular ornamental shrub Japanese quince (*Chaenomeles*, which has rather tasteless fruit. Quince used to be widely used but has now fallen out of favour. The tree is actually very pretty and long lived, and suitable for small gardens. The fruit can be used in jams and jellies.

FLOWER: The quince has glorious rose pink blossom.

FRUIT: The acid fruit looks like a small yellow pear, has a very strong fragrance and should be gathered in October before night frosts. Fruit will continue to ripen when stored and will keep for about six weeks. The late Christopher Lloyd says it should never be stored near apples or pears as it can affect their flavour. He also advises not bothering to peel the fruit before cooking as it virtually melts away in the pan.

PRUNING AND CARE: Quince, like so many other fruit trees, likes to be in a warm and sheltered spot, but it also likes to be well watered, so is often planted near ponds. Plant out in October or November. Little pruning is required, thin out branches to let in light and air.

VARIETIES: Lusitanica (formerly known as Portugal) has large mildly flavoured fruit and is very vigorous.

FIG *(Ficus carica)*

DESCRIPTION: Figs are native Mediterranean plants but they can be grown in the south and west; further north, however, you are unlikely to get any fruit unless figs are grown under glass. Figs are attractive garden plants, often grown for their foliage alone, but they can grow very big if left untended and if grown for fruit then the roots must be contained.

FLOWER: Insignificant

FRUIT: Fruit should be left to ripen on the tree and is usually ready for picking between mid August and October, depending on conditions. It is ripe when the stalk softens, the fruit hangs down and the skin begins to split. Before consuming it is often best left to ripen for a day or two after picking. Figs can be dried or bottled, but are best eaten fresh. Remove any fruit larger than a pea at the end of the summer. Tiny fruit can be left on the plant as they can survive the winter and start to grow the following year.

PRUNING AND CARE: If the roots of a fig are not contained, the plant will produce too much growth and will not fruit. It is best grown against a sunny wall with the wall restraining its growth on one side. Dig a one metre square hole, line it with bricks and paving slabs to a depth of 30 centimetres. Fill the hole with a mix of garden soil and stones. Figs are pruned in winter; strong wood is the most fruitful, green shoots should be removed. Fan-trained trees should be supported with wire. Figs can also be grown in containers.

VARIETIES: Brown Turkey is one of the best-known varieties.

STRAWBERRY (*Fragaria*)

DESCRIPTION: Strawberries are my favourite fruit of all. As a child, I just couldn't get enough of them and, with all my key family events happening in June and July, I also associate them with good times. However, not all strawberries are so delicious. Increasingly supermarket strawberries are partially green, hard and tasteless. Strawberries bought from local grocers and farmshops are infinitely better. Similarly, if you find a good Pick-Your-Own (PYO) site, check out which variety they are growing, as even PYO can be disappointing. This is one fruit that should not be eaten out of season: it is worth the ten month wait for June to come around and then you can gorge yourself. Growing your own fruit takes this delicacy onto a new level of taste experience. Strawberries that are not being grown commercially and picked fresh are ripe balls of juice. They take a little care to grow and require a little bit of space, but the fruit will be worth all the effort. Alpine strawberries are fabulous too, but they only carry a little fruit. Bob Flowerdew advises that if you pick them as they ripen and immediately freeze them, in time you will have enough fruit to make a heavenly jam.

FLOWER: Delicate white flowers appear in May, June and July. If late frosts are threatened, protect the flowers by covering them with old newspapers at night.

FRUIT: If you plant strawberries in late summer you will have a crop the following year, but if you plant in March remove all flowers in the first year and wait for a crop the following year. A four metre long row of plants will keep you and the family well supplied with berries! The bright fruit appears from May to October, depending on the variety you grow. You will need to either net your plants or protect them with cloches – otherwise the birds will enjoy the fruits of your labours.

PRUNING AND CARE: Strawberry plants are most prolific in their first two years of life and after three years they should be discarded. However, plants will throw out rooted runners which you can pot up in tiny pots and leave in situ until the plant is fully rooted, when it can be severed from its parent. This will give you a steady supply of fresh plants. To keep a constant supply of fruit, you should replace one-third of your row of plants annually. Strawberries can be planted in a sunny position in September or March, spaced at 45 centimetres intervals in rows that are 75 centimetres apart. The fruit needs rich soil with farmyard manure or compost dug in. Do not allow weeds to develop between your rows. Apply an annual mulch of manure in May and run straw between both plants and rows; this helps to keep the fruit away from the soil and minimises the risk of infection. In the plant's second year, allow each plant to develop one runner; you can help rooting by pegging this into the soil. Plants will need watering when the weather is dry, but spray between the rows rather than the plant itself as the fruit is so easily damaged by water, and pick and remove any fruits that are rotting or mildewed. Remove or burn straw in autumn to help get rid of any lingering pests. Plants grown in containers will need a lot of water.

VARIETIES: It is worth mixing the varieties as this extends the fruiting season and if you have a wet few weeks, you won't lose your entire crop to mould. The pre-war variety Royal Sovereign is unbeatable for flavour, but doesn't supply a heavy crop and is susceptible to disease; Grandee is also a superbly flavoured fruit. Perpetual varieties are slightly less sweet, but keep the fruit supply coming for three months or more, Gento will produce from June to October and St Claude from July to October. Baron Solemacher is a good alpine strawberry.

APPLE *(Malus domestica)*

DESCRIPTION: The apple, with a 4,000 year history, is one of the oldest fruits. It grew wild for many years before being cultivated. Remains have been found preserved in prehistoric sites. The apple is the third most produced fruit in the world, losing out only to oranges and bananas. The UK apple market is worth £320 million, but only 30% of the fruit sold on the home market is produced here. Additionally, the UK is the only country to grow apples especially for cooking. This fruit is one of the easiest to grow and one apple tree alone can give a very satisfactory crop allowing you to bake Dutch apple cake, apple charlotte, tarte tatin and apple strüdel as well as apple sauce and apple jelly.

FLOWER: Apple blossom appears in May, when trees are smothered in beautiful pink white flowers.

FRUIT: An established, well-cared for apple tree can produce an impressive 35–45 kilos of fruit. Apples are ready to pick when they lift easily from the tree, stalk and all, with a slight twist. The fruit is best stored wrapped in paper, or you can store a number of apples together in ventilated plastic bags.

PRUNING AND CARE: Apples appreciate a sunny sheltered site, but they will produce fruit over most of the United Kingdom.

Frost pockets can damage the spring blossom and reduce your crop, so take care to avoid cold corners. Plant out between November and March, though avoid doing so if hard frosts are forecast. Trees can be purchased from one year old, though they will only have a single leader. If you want to make life easier buy a tree that is two to three years old – the nurseryman will have created a good growing structure for you. You can also grow apples as espaliers, step-overs (a one storey espalier) that can be used to screen a vegetable garden, or as cordons. There are bush trees, standards, pyramids and dwarf stock – the choice is daunting. Discuss your space with a nursery specialist and they will guide you towards the most suitable form – from there on in you simply have to select your variety! Once the tree has begun to fruit – usually within two to five years depending on the vigour of the rootstock – you should remove any damaged, diseased or crossing branches and similarly destroy suckers when the tree is dormant (November to February). Apple trees are pruned according to their fruiting pattern. Spur-bearing plants need branches to be shortened to promote fruiting. Tip-bearing fruit trees are more tolerant of neglect; they fruit from the top bud of branches. It is important to thin your crop in June as too much fruit will strain the tree. Use scissors to thin and at the same time remove damaged or diseased fruit. Dessert apples should be thinned to 100–150 centimetres apart, while cooking apples should be 150–220 centimetres apart. After harvest, all remaining apples should be removed from both tree and ground to help prevent the spread of disease. Plant chives, allium, nasturtium and penstemon nearby as companion plants, to help keep sawfly and woolly aphids away. Never plant an apple tree where one has been before.

VARIETIES: There are more than 2,000 varieties of apple: supermarkets commonly stock eight of these – so you are spoilt for choice! There are crab apples, dessert apples, cooking apples or cider apples to choose from, so make sure you purchase a variety that suits your requirements. Blenheim Orange is a good dessert variety to grow; it has been around for more than 200 years and has a good reputation.

MULBERRY (*Morus nigra*)

DESCRIPTION: This is one fruit that you simply cannot buy commercially – the only way to taste it is to grow it yourself. It is an attractive, slow-growing tree with heart-shaped leaves. It will, ultimately, become very large – so you do need the space to accommodate it. Patience is also required as it is unlikely to produce fruit reliably for the first seven years; however it is worth the wait! The further north you go, the less likely you are to obtain fruit.

FLOWER: Insignificant

FRUIT: Mulberries produce dark red fruit in August and September; they are delicious, but worse than blackberries for staining the hand. The simplest way to collect them is to put a clean, but old, sheet underneath the tree and then shake the branches – picking them individually by hand is time-consuming and they easily squish between the fingers. Birds will take a lot of the fruit, but there is not much that can be done about this when the tree gets really big. Fruit can be bottled or frozen and it makes an excellent jam or jelly.

PRUNING AND CARE: Plant in a sunny spot between November and March. Mulberries do not need any real attention; in fact they do not enjoy being pruned. However, branches are inclined to droop as they age and may require support from wooden crutches. Remove old dead wood, or young branches growing inward. The tree is self-pollinating.

VARIETIES: Chelsea bears fruit earlier than most other varieties. *Morus alba*, the white mulberry, is nothing like as attractive and produces tasteless fruit. It is commonly grown to supply silkworms with their required diet of mulberry leaves.

APRICOT *(Prunus armeniaca)*

DESCRIPTION: Home-grown apricots taste divine. They are quite easy to grow, but harder to get to fruit. They require a warm, sheltered spot, ideally against a sunny wall and well-drained soil. They will fruit grown out of doors in the south, but in Wales and the north you would be better advised to grow them in a greenhouse or conservatory. Trees do not usually fruit until they are four years old, so the age of the tree you purchase is significant, plant out from November to March. Of course you can always try growing your own from the stone, which is quite easy to do.

FLOWER: Apricots produce glorious white flowers flushed with pink in February. This timing, when it is still cold and frosts are common, damages the flowers and therefore affects the size of the crop. The early flowering also means that there are far fewer insects around and therefore pollination is also a problem. The best solution is to take a fine paintbrush and pollinate the blossom by hand.

FRUIT: The plump orange fruit should be thinned in April or May when it has grown to about the size of a gooseberry. If you don't do this your whole crop will suffer. Thin until fruit is at five-seven cm intervals. Birds love apricots so you would be advised to net to protect your crop. This plant can be cropped from late July through to August. Do not attempt to pick the fruit too early: if it doesn't come away in your hand then it is not fully ripe – a good indicator is to wait until some of the fruit has started to fall naturally. The fruit will only last a few weeks fresh, but it can be bottled, frozen or dried. Jam made from home-grown apricots is excellent. Trees will bear as much as 40 lb of fruit.

PRUNING AND CARE: Ensure that drainage is good before you plant. Mulch the tree every winter and water regularly in very dry weather, particularly when it is fruiting. Pruning is not complicated because the plant will fruit on wood over one year old as well as on older wood as well, however as with all fruit trees it is important to keep the structure open.

VARIETIES: One of the most reliable and popular varieties is Moorpark, but Bredase is good for small gardens.

CHERRY *(Prunus* cerasus*)*

DESCRIPTION: Cherries are quite fulsome trees and you will need at least two for cross-pollination if they are to fruit. They like well-drained soil and will not tolerate heavy clay. The fruit comes with a sweet and sour option, the sour being the Morello cherry which thrives grown against a cold north wall and which will produce perfect fruit for culinary use and will make magnificent jam. Sweet cherries can be grown as trees, bushes or against walls, but you will be fighting a running battle with the birds, they are less keen on Morello cherries but these plants will require careful annual pruning.

FLOWER: Cherry trees are smothered in the most beautiful white blossom in spring and make glorious specimen trees planted in grass.

FRUIT: Sweet cherries fruit in June and July, sour cherries in July and August. Cut them off with scissors and not by hand to avoid damaging the wood and making it susceptible to disease. Birds love the sweet cherries and will consume your whole crop if given the chance – plants need to be netted. Cherries do not keep well and will only last for about two days if kept refrigerated.

PRUNING AND CARE: Plant in well-drained soil and stake the tree to prevent root damage from rocking. Sweet cherries do not need heavy pruning, though any damaged wood should be promptly removed to prevent disease; cherries are very prone to Silver Leaf disease. Sour cherries are produced on wood grown in the previous year and therefore should be fed and limed in spring to promote growth and pruned in spring to produce new shoots. Fan-trained trees, grown against walls, will need to have outward growing shoots removed annually. Unproductive branches should be removed every three years or so.

VARIETIES: The Morello cherry is the best sour fruit. May Duke and Sunburst are self fertile sweet cherries, Frogmore Early will give you a yellow cherry mottled with red, but will need to be cross-pollinated with Late Amber and Early Rivers will give a dark purple cherry, but it will need to be cross-pollinated with Merton Bigarreau.

PLUM (*Prunus domestica*)

DESCRIPTION: Plums are a luscious dessert fruit, wonderful baked or stewed and make a divine jam which keeps for ages. The plum is one of my favourite fruits to grow, it is tolerant of neglect and so obliging in terms of fruit production. Gages, a less robust plum, have a wonderfully delicate and slightly more acidic flavour, but have all but disappeared from the shops.

FLOWER: Pinkish white blossom appears in spring.

FRUIT: Plums and gages fruit in July and August and appear in a fabulous range of colours from yellow, through peachy pink and on to dark purple. A fully-grown plum tree can produce a whopping 25–40 kilos of fruit, depending on the variety and growing conditions, so one tree is usually sufficient. However, take care to purchase a self-pollinating variety if you are just having the one tree. When picking plums the stalk should come away with the fruit. Leave on the tree to ripen, but if you are using them for cooking or for jam. pick them just before they are perfectly ripe. Gages should be picked before they are perfectly ripe or their skins will split. Birds and wasps will go after your crop; you can protect fruit by putting it in bags on the tree!

PRUNING AND CARE: Plums do best in a sunny sheltered spot in soil that is neither too dry nor too waterlogged. They should be planted from November to March, the earlier the better. Feed annually with a mulch of farmyard manure for best results. Plums fruit on old wood as well as the previous season's growth. Fruit may need thinning, it will drop naturally in June, after which time you should only have one plum every 50–75 millimetres. Trees are usually sold at two to three years old and should already have a structure established, from then on in you really can leave them alone, you will only need to take out dead wood or crossing shoots, do this immediately after you have picked your crop. To train a fan-trained tree, prune it as you would the peach for the first three years after planting. Pruning thereafter is about maintaining the fan shape of the tree; in spring remove inward and outward growing buds in July, pinch out side shoots that are not needed as branches and, after cropping, shorten them again by half. Plums are susceptible to Silver Leaf and this will enter via any pruning wounds made in autumn and winter, you may wish to paint the cuts with tree paint. Take care when weeding or mowing around the tree – the roots are easily damaged and it will respond by throwing up suckers that need to be pulled off.

VARIETIES: Victoria is one of the best known varieties; it is a juicy, warm pink fruit and is self-pollinating, although if you can house two trees then choose a variety compatible with Victoria, such as Early Rivers, to encourage cross pollination. Severn Cross is a self-fertile greengage.

DAMSON (*Prunus domestica insititia*)

DESCRIPTION: For me, the total bliss of damsons is picking them from the hedgerow. I return from every short walk with pockets bulging. I am lucky that, in Kent, the hedgerows are groaning with fruit in August and September. Damsons have an intense blue bloom which on cooking releases the darkest purple juice. Few jams can compare with damson, although it is nigh on impossible to get rid of every stone, but damsons mixed in with fruit crumbles, particularly apple crumble, turns an ordinary dessert into a gourmet's delight. If you don't have any damsons around you try putting a few into a native hedge or you can grow them as a bush or fan train them against a wall.

FLOWER: Small white flowers in April.

FRUIT: Fruit appears in August and September.

PRUNING AND CARE: Fruit grows on old and new wood, little pruning required, but it will cope with harsh annual trims as part of a hedge.

VARIETIES: Merryweather produces good fruit for desserts and is self-pollinating.

PEACH AND NECTARINE (*Prunus persica*)

DESCRIPTION: Peaches have been grown in Britain since the Romans came to visit and they are not hard to grow in a sunny, sheltered spot. The proper fan-trained tree, grown against a south-facing wall, is a thing of beauty, but you will have to commit yourself to some serious annual pruning and tying. If you can find the time, it will pay handsome dividends; a home-grown peach, with its incredibly juicy white flesh is a true delight. Nectarines are essentially tender peaches, without the velvety skin but they are nigh on impossible to grow out of doors in the UK and the yield, when they do fruit, will be far less than that from a peach tree.

FLOWER: Peaches have beautiful pink flowers in the spring, the trees are worth growing for the blossom alone.

FRUIT: Fruit will ripen as early as July if you live in the sunny south, but trees in the north can be cropped in September.

PRUNING AND CARE: Peaches should be planted in full sun, in well-drained soil between October and January, one tree is usually sufficient to supply a family with enough fruit. If you are training against a wall or fence, use vine eyes to secure wire supports 150mm apart. Peaches need to be pruned twice each year once they are two years old – in February and again

Bottled fruit looks and tastes divine

in late summer after harvest. In February, cut about one-third off each main branch and after harvesting cut out the fruiting lateral. When the tree is in bloom, remove any shoots growing inwards or outwards then pinch out the growth buds on flowering laterals leaving three – one at the base, one in the middle and one at the top. The ground must be kept moist in the growing season. It may be necessary to give nature a helping hand with the pollination process with a child's paintbrush. Fruit should be thinned gradually until they reach the size of a golf ball by which time you should have a single fruit every 23 centimetres.

VARIETIES: Peregrin is the best-known and most reliable peach variety, but Lord Napier and Rochester are also a very good choice.

SLOE (*Prunus spinosa*)

DESCRIPTION: Sloes are a native hedgerow plant, better known as whitethorn because of their dark thorny wood. It has a delicate blossom and sour, deep blue fruit. Mix plants with hawthorn, wild rose and holly to make a native hedge. Use fruit to make sloe gin, see below, or make a sharp jelly for use with meat.

FLOWER: The delicate white flowers are the first blossom to appear in late February or early March, an indicator that spring is on the way.

FRUIT: The blue black fruit should not be picked until after the first frost has occurred. Sloe gin is delicious – make it straight after picking. Remove half the contents of a gin bottle, then add an equal weight of sugar and fruit, pricking each sloe with a fork to release the juice (time-consuming but it makes all the difference). Shake it regularly and keep until Christmas before consuming.

PRUNING AND CARE: This is a tough old plant that needs little or no attention and which will withstand heavy pruning – hence it's success as a hedging plant.

VARIETIES: This is a native plant with just the one strong variety.

PEAR *(Pyrus communis)*

DESCRIPTION: A good pear is thin-skinned, with soft, juicy, slippery flesh and an intense, but delicate flavour. It's a perfect tree for a small garden with beautiful fragrant blossom. You can prune it to keep it small and make it easy to crop. Pears do need a lot of moisture and are not good grown in grass as it competes for moisture. However they can be grown as espalier screens or fan trained against a wall. Many varieties need cross pollination so you will need two trees of a compatible variety. Pears are long lived and will fruit successfully as long as they are given some regular pruning and feeding.

FLOWER: Fragrant greenish white flowers in early spring.

FRUIT: Pears are harvested between August and October, depending on the variety and the location in which it is grown. The fruit can be picked when it parts easily from the tree retaining its stalk. An established tree can produce as much as 22 kilos of fruit, an espalier 12 kilos and a cordon about 2.5 kilos. If fruit is not quite ripe on picking, store carefully and ensure that the fruit does not touch each other.

PRUNING AND CARE: Pears need to be planted in a sunny, sheltered position in good, rich, well drained soil (they don't like clay or sandy soils) ideally in November, but planting can take place until March. The tree should be staked and tied. Trees should be planted a minimum of four metres apart. A mulch of good farmyard manure each January or February is helpful. Most pears fruit on spurs therefore they need branches to be shortened to promote fruiting. Tip-bearing trees fruit from the top bud of branches. Older pear trees can be pruned harder than apple trees; in winter remove any dead twiglets or ones with only a few fruit buds. Thin and keep the centre of the tree open.

VARIETIES: Doyenne du Comice is a wonderful fruit if grown in a warm and sheltered situation, it needs to be cross-pollinated with Conference which produces a generous crop, or Williams Bon Chrétien (a tip bearer) which is a truly delicious fruit.

CURRANT (*Ribes*)

Description: Blackcurrants (*Ribes nigrum*) and Redcurrants (*Ribes rubrum*) are deliciously easy to grow and fantastically rich in vitamin C – so they are both delicious and nutritious! Children often don't like them, finding them too acidic, but as their taste buds mature they'll will find them irresistible in blackcurrant tart, summer puddings, a deep pink blackcurrant ice-cream or a rich redcurrant jelly. Bushes need to be netted to protect the currants from the birds.

FLOWER: Small insignificant flowers clustered at the base of shoots and spurs.

FRUIT: Fruit appears from June until late October in a good year. The fruit freezes easily. A mature blackcurrant bush will produce around 5–7 kilos of fruit. A mature redcurrant bush will produce around 5 kilos of fruit.

PRUNING AND CARE: Plant in October to March. Blackcurrants in particular don't like to be in dry soil. Dig over the site and manure well, then plant bushes 1.5 metres apart. After planting cut all blackcurrant shoots back to the base to two buds to encourage new shoots to spring up. From then on, cut out old fruiting wood and allow new shoots to develop. Bushes last for up to about eight years, but should then be replaced. On planting redcurrants cut back stems to four buds above the base, the topmost bud should be pointing outwards. Unlike blackcurrants, redcurrants flower on old wood, so pruning consists of cutting back the leader shoots on the main stems, cutting out all weak and broken stems, then cutting back the side shoots on the remaining stems. Aim for the centre of the bush to be nice and open to help the ripening process with around six or seven main branches in total. Redcurrants can also be grown on wire as cordons.

VARIETIES: Many of the newer varieties of blackcurrant have been developed to travel well. For juicy thin-skinned fruit, older varieties such as Boskoop Giant are best. Ben Nevis is a newer variety with good disease resistance and fruits later.

Raby Castle is a good redcurrant bush, which produces particularly large fruit.

GOOSEBERRY (*Ribes* uva-crispa var. *reclinatum*)

DESCRIPTION: Gooseberries are the most deliciously tart fruit and they are best when grown and picked from the garden. Shop-bought produce tends to be thick-skinned, unpleasantly acid and somehow not the same plump, juicy fruit at all. Moreover, not all gooseberries are green, as in the supermarket. The fruit can be flushed yellow, red or white and each colour has a particular flavour. The best news is that they are very easy to grow, so long as the plant is fed and pruned, the fruit freezes well and makes wonderful jam or a sharp sauce.

FLOWER: Insignificant

FRUIT: Individual berries can weigh an astonishing 50 grams and if different varieties are planted you can have fruit from May to August. Some varieties are sweeter and can be eaten as dessert fruit; others are sharper and are best cooked. The smaller the fruit the more acid it will be and the skin will be softer. One bush will produce approximately 4–5 kilos of fruit. Birds will eat gooseberries, so protect with nets.

PRUNING AND CARE: Gooseberries will grow in sun or semi-shade but they do not like late frosts. They are best planted in October when the soil is still warm, but can be planted out from then until March. Dig a hole and fill with good soil mixed with plenty of well-rotted manure. Bushes should be spaced 1.5 metres apart. Give them a top dressing of manure in spring and then thin fruit in May so that there are 75 millimetres between each individual fruit. Prune annually in winter to create a goblet-shaped bush that is open in the centre; remove crossing and inward-growing branches, or sagging exterior branches. Caterpillars will attack this plant, the best organic solution is to shake them off and remove.

VARIETIES: Leveller can be used as both a dessert and a culinary fruit. May Duke can be picked early for culinary purposes or left to turn red in July for a dessert fruit. Early Sulphur is an old hairy yellow variety – a gooseberry as it used to be, plus it has the advantage of being one of the earliest varieties to fruit.

RHUBARB *(Rheum* x *hybridum)*

DESCRIPTION: Rhubarb is a slightly tart pink fruit (technically a vegetable), which can be forced into readiness as early as January. Once you have a plant or two, you have a regular supply of this delicious fruit for many, many years to come. Forced rhubarb is even sweeter and tenderer than the first pulls in June. It is wonderful stewed with custard, the kind of good old-fashioned dessert that I was raised on, or baked in the oven with ginger.

FLOWER: Rhubarb has rather splendid white flowers and beautiful seed heads; some people advise that they are removed, but I personally like a plant that both looks good and tastes delicious so am inclined to leave them alone.

FRUIT: Three plants will supply the family with plenty of fruit, but just one plant will allow you to have the wonderful taste of your own fresh rhubarb. Do not eat stems in the first year and then pick sparingly in the next two to three years. Rhubarb is tender for six weeks on ripening, after that it gets a bit stringy. Pull stems by twisting rather than cutting. Please remember that whilst the pink stems are divine, the leaves are poisonous – put them on the compost heap!

PRUNING: Rhubarb is delightfully easy to grow, its only preference is for an open situation. Dig in manure a month or so before planting, either in March or October and then give it an annual mulch of manure in spring. Rhubarb can be forced when it is three years old, lift one plant and turn it over to expose the roots to frost – this pushes the plant into dormancy early. In December lift the plant, put into a large pot or box and put into a greenhouse or shed. The rhubarb must be in darkness so you may need to be creative with some black bin bags. Rhubarb can also be forced in the garden by covering the crown in early February with a large bucket or box – put straw around this for further protection. Indoor forced plants can be replanted but will take a while to recover.

VARIETIES: Victoria is a reliable variety and Timperley Early is good for forcing.

BLACKBERRY *(Rubus fruticosus)*

DESCRIPTION: Everyone knows this fruit and the prickles that come with it, mostly from gathering it wild. It's incredibly easy to grow and will be covered in fruit, the downside is that it is a big plant that can get out of hand. On the plus side, it can make life very difficult for any would-be intruders when grown around the boundary. This plant is best grown by blackberry fans who do not live anywhere near fruit-bearing wild hedgerows.

FLOWER: Flowers are white flushed with pink and appear from May through to September.

FRUIT: Plants will produce the glossy black fruit from July to October. Blackberries do not keep for long, so pick only what you need. Bramble jelly is the most perfectly delicious dark purple conserve.

PRUNING AND CARE: Plant blackberries in November in a sunny position and then cut canes back to around 25 centimetres from the base. As the shoots grow, tie them back, partially to limit the spread of the prickles. Canes fruit best when they are one season old and the simplest way to keep control is to tie all one season's growth in one direction one year and then, in the next year, direct the new shoots the opposite way.

VARIETIES: Bedford Giant is a good strong garden variety, while Oregon Thornless has really beautiful leaves and no nasty prickles.

RASPBERRY (*Rubus idaeus*)

DESCRIPTION: Raspberry is the most intensely flavoured fruit and one of the easiest soft fruit to grow, which, given the price and the minuscule portion size of shop-bought fruit, makes it all the more extraordinary they are not more commonly grown. Raspberries are delicious to eat either on their own, or more decadently, sweetened with sugar and combined with cream. They also make fabulously tangy dessert purées, especially when teamed with sweet peaches or meringues.

FLOWER: The plant produces small white flowers.

FRUIT: Raspberries come in a warm spectrum of colours: yellow, orange, pink, red and deep red, depending on the variety selected. Fruit is ripe when it pulls easily away from its stem. Bushes will produce fruit for up to 12 years, after which they should be replaced. Ten canes will produce about five kilos of fruit. Unlike strawberries, raspberries freeze very well, if the birds don't get to them first – you will have to net your canes.

PRUNING AND CARE: The plant thrives in cool damp conditions, which goes some way to explaining why they grow so very well in Scotland. Raspberries grow on the previous year's canes, therefore annual pruning is essential, although autumn raspberries are produced on the current season's growth. Raspberries should ideally be planted in November or December in a warm situation, though the plant will tolerate partial shade. Dig in some farmyard manure a month or so before planting and then continue to feed with compost or a general fertiliser every spring. Plant canes approximately 50 centimetres apart and keep a distance of 1.8 metre between rows. Cut the old cane down to ground level when the new growth appears in spring. When canes have fruited cut them down, retaining six to eight of the young canes that have not yet borne fruit. These should be tied onto wire supports. In February, check that no canes are growing above the top of the wires.

VARIETIES: Summer and autumn fruiting varieties are available. Glen Ample, a summer variety, crops well and Golden Everest is a sweet autumn plant.

ELDERBERRY *(Sambucus nigra)*

DESCRIPTION: This is a plant with numerous uses. The flowers can be made into cordial and fritters and the fruit used for cordial and wine. Again, like the damson, the countryside is full of elderberries and there is little need to cultivate your own.

FLOWER: The flowers appear from mid May to June, if you want to use the flowers for fritters or cordial they must be picked when in full flower, but before the petals start to fall. If you pick when the flower head is in bud you are picking too early – it is a fine art and flowers are only perfect for picking for a day or two.

FRUIT: Elderberries can be gathered from mid August to October.

PRUNING AND CARE: Check growth as and when necessary and if flowering starts to decrease cut the wood back hard; you will lose flowers and fruit for a year but the tree will come back as vigorous as ever.

VARIETIES: Laciniata has beautiful creamy flowers and will grace any garden, Guincho Purple has pale pink flowers and purple foliage.

GRAPE (*Vitis vinifera*)

DESCRIPTION: Grapevines are beautiful plants worth cultivating for their looks alone. They are ornamental climbers, with beautiful foliage that turns bright red in autumn. The Romans are said to have imported the grape vine to the UK before colder weather made it impossible to grow them. For many years, only the best households ate grapes, growing them under glass in specialist grape houses, which would produce fruit from June until January. The process was time consuming with flowers being hand-pollinated and bunches of grapes individually thinned. As our summers have got warmer, grapevines are now doing quite well in the United Kingdom, notably in the south and west. Vineyards are springing up and British winemakers are finally producing a quality product. Last summer, I was repeatedly invited to pick bunches of dessert grapes, direct from the vine, at various friends' houses. The fruit was delicious even though the grapes were tiny, doubtless because no thinning had been done. The longer, warmer summers are enabling many of us to enjoy eating grapes direct from the garden for the first time. A heated south-facing conservatory will enable you to grow indoor varieties in a container, though you will have to help them along with some hand-pollination, feeding and pruning.

FLOWER: Insignificant

FRUIT: There are distinct varieties of grape that can be grown out of doors and others for cultivation under glass. White outdoor varieties of grape tend to fare better than the red outdoor varieties. Early plants will fruit outdoors in September and October, indoor varieties from September. An established vine will produce about 2.5 kilos of fruit a year. Birds and wasps will take your fruit if given the chance; nets will help keep them away but can make the fruit prone to mildew. If frosts are imminent, you can pick grapes on a long stalk and then store them in a bottle of water in a cool place to continue ripening.

PRUNING AND CARE: Grapes should ideally be grown against a warm wall, preferably south-facing and should be planted in well-drained soil. However vines can fruit on sunny pergolas and arches. Plant out from October to March, but March is the ideal month for planting. Put a stout two metre stake in to give the plant support and space plants at 1.5 metre intervals. In the first year, only allow the strongest shoot to grow up the support, pinch out other shoots at two or three leaves. However, grapes do seem to withstand some healthy neglect if you just want to try your hand. When you can see bunches

Thin grapes to two to three trusses per lateral

forming, thin them out so that you have one bunch every 60 centimetres; and you will get a much better crop. It is horribly tempting to leave them, but the end result will be more than worthwhile if you bite the bullet and thin. Growing grapes indoors is more complex. The previous summer's growth must be pruned back to one bud in November/December. At the same time, the glass should be washed with disinfectant. Ventilation must be good between January and March to keep plants dormant, but temperatures increased in April to promote growth and maintained until flowers have set. Plants must be top dressed with fertiliser and manure every spring.

VARIETIES: Boskoop Glory is a good outside black dessert grape and Madeleine Silvaner a good outside white for both wine and dessert. Schiava Grosser (formerly known as Black Hamburgh) is a reliable black grape to grow under glass and Foster's Seedling is also an excellent white.

TOOLS AND EQUIPMENT

Very few tools are needed to grow fruit. A lot of my tools have been inherited from family and I am quite sentimental about them, so they are only replaced when they break. I am also drawn to old garden equipment, which can be found in antique shops or at jumble sales and boot fairs. Every so often, they are replaced with a superb piece of modern design that utilises the latest high tech materials. If you can afford it, it is worth investing in decent tools, as in this instance you really do get what you pay for.

IT WILL MAKE YOUR WORK EASIER if you keep your tools in good condition. Scrape earth off spades, forks and hoes before putting them away; don't leave wet vegetation in your wheelbarrow and make sure that secateurs, hoes and saws are all kept really sharp.

SPADE: This is essential for digging. Spades come in various sizes, make sure you select one you are comfortable with. I prefer to use a very small spade, I find it easier to work and it is kinder on my back. Forks are used to break up lumps in soil and are useful when working a compost heap, never use them to lever out anything really tough, such as an old tree root, as the tines can bend.

HOE: A good hoe is well worth the money. There are two basic kinds: draw hoes, which chop out weeds (a short-handled version is available for use when working on hands and knees) and a Dutch hoe or patent hoe which will to help keep a patch of land weed free once cleared. I would not now be

parted from my rubber-handled Dutch hoe, which is terribly
gentle on the hands!

SECATEURS: A good pair of secateurs is an abolute must for
pruning fruit trees. Keep them sharp so that they will cut
clean and make light work of this task. Long-handled

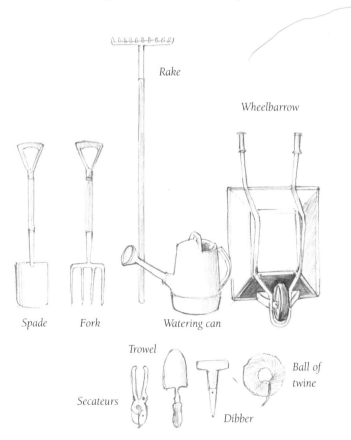

Rake

Wheelbarrow

Spade   Fork   Watering can

Trowel

Secateurs

Dibber

Ball of
twine

secateurs are useful for pruning higher up the tree – they are more powerful than the regular variety and can cut particularly large branches. A long-handled pruner will enable you to prune tall leaders from the ground – this has a reach of around 350 centimetres and is operated by a handle at the base of the tool. A budding knife is also useful for making small, precise pruning cuts for buds and young shoots.

SAW: A pruning saw, which is curved and narrow at its tip, makes light work of cutting small branches up to about 10 centimetres in diameter.

WHEELBARROW: A wheelbarrow is an investment and will last a lifetime as long as you treat it with respect and don't use it as a fireplace! The chore in gardening is not usually the work itself, but the clearing up afterwards. If you fill the wheelbarrow as you work and empty it regularly, you will clear up as you go. It will also be useful for the enjoyable and rewardig task of carting your bumper fruit crops from garden to kitchen door.

WATERING CAN: A watering can and a hose are also required. You will need to water young plants in their first summer, or some plants when they are fruiting. Always get a control handle for your hose, even if a hose pipe ban is in place and you can't use water liberally, you will still be able to take the hose to your fruit garden and fill your watering can in situ to make the task less arduous.

COMMON PROBLEMS

If you take reasonably good care of your plants, opt for sensible sturdy varieties, and plant them in the appropriate situation, at the right time of year, in fully prepared soil, you will have ensured that your trees and plants have the best start, that they will grow strong and vigorous with the ability to resist many diseases that a weaker plant might succumb to. Many problems are avoided by simply sticking to basic rules and by not trying to bend nature to your will.

SIMPLY KEEPING AN EYE ON THE HEALTH of your plants can help you to eradicate any local difficulty before it has an impact. If you prune out any damaged or diseased wood and ensure that your gardening equipment is sharp so that pruning cuts are clean, your trees are less likely to fall prey to a fungal disease or virus. If you watch out for holes appearing in leaves, or curling and browning leaves, you may escape a mammoth infestation if you simply pick off the affected parts and burn them. Blackcurrants are very prone to big bud; an insect crawls into the bud and it becomes swollen, the solution is simply to pick off any affected buds.

Similarly, if plants look as though they are being eaten, then inspect for caterpillars and dispose of what you find (the organic gardener Bob Flowerdew recommends dispatching them to the bonfire but I prefer feeding them to my chickens). Rotting and mildewed fruit should be removed from the plant promptly and burned before it gets the chance to spread.

Barrier methods are a simple way of keeping many pests away from your crop – nets and wire will keep out birds, butterflies, moths, rabbits, squirrels and rodents – even slugs can be deterred by circling plants with materials they will not cross. Humane traps will catch mice. Old carpets laid around gooseberries, pear and raspberries can trap the pupae of specific troublesome flies and beetles on hibernation. Grease bands or sticky bands tied around the trunks of fruit trees will prevent crawling insects from reaching their destination. Trap slugs and snails with saucers of overripe fruit and then dispose of them a good way away.

If you follow the principles of organic gardening and actively encourage natural predators into your garden – hoverflies, ladybirds, lacewings and assorted beetles, for example – they will take care of many of your pests for you. Open-centred flowers, such as marigolds (*Calendula*), nasturtiums (*Tropaeoleum*) fennel (*Foeniculum*), the cornflower (*Echinea*) and poppies (*Papaver*), all attract hoverflies into the garden. Build winter shelters for lacewings with bundles of hollow wood such as bamboo canes – these are available from garden centres if you don't want to make your own. Plant some evergreens in your garden to provide further winter shelter for predators.

An insecticide can be made from the herb pyrethrum (*Tanacetum cinerariifolium*), which can be used against aphids and spider mites. It can be used on the skin as an insect repellent and can be sprayed on insects that invade the house: ants, flies and mosquitoes. The plant has been used as an insecticide for centuries – the active ingredient within it paralyses the nervous system of many insects, but is non-toxic to mammals and environmentally friendly. However it will still kill beneficial insects.

The herb horsetail (*Equisetum arvense*), commonly regarded as a weed, can help to combat root or fungus diseases. Make up a solution of one part horsetail to 50 parts water – do not make it any stronger as horsetail is very powerful. Boil the mix for 15 minutes and allow it to cool. The same day check that the soil around affected plants is not too dry; if it is, water it well. Apply the mixture the following day around the base of the affected plants; do not spray it onto the leaves.

Plant suitable companion plants close to your trees to help them combat pests. These work in various ways: by attracting predators to the area so that they can kill the pests; by attracting pests away from your fruiting plant; or by improving a fruiting plant's resistance. Chives (*Allium schoenoprasum*) and garlic (*Allium sativum*) are beneficial to apple trees and chamomile (*Chamaemelum nobile*) is believed to act as a general tonic, restoring the health and vigour of nearby plants. Plant pyrethrum close to susceptible crops to help keep insects away.

Wrap fruit in paper before storing it

Generally speaking, I do not believe in using chemicals to treat problems. Commercially grown apples and pears are automatically sprayed several times each year to eradicate assorted pests and diseases on a better-to-be-safe-than-sorry basis. Part of the attraction of growing my own fruit is that it gives me a small measure of quality control in at least one area of food production. If, however, you want to use chemical fungicides and pesticides, they are easily available from most garden centres.

Browning leaves

This is often a sign that the plant has been hit by bacterial disease or a virus. Prevention is the best cure, follow planting procedures to the letter and you should ensure no problems arise. However, if you do notice a problem try to deal with it by removing the affected parts of the plant, Pests can also be responsible, so check for caterpillars.

Rusty leaves

Rusts will often attack a number of plants of the same variety simultaneously in damp conditions. Leaves turn yellow and raised brown spots appear; remove the affected leaves and burn them and also remove any decaying leaves around the plant.

Mildewed leaves or fruit

Gooseberry is particularly susceptible to mildew, which is at its worst in warm, damp weather. Remove all diseased shoot tips. If apple or pear leaves, shoots or flowers, develop a white

Blackcurrants: remove old wood and open the centre

Blackberries: remove shoots that have fruited in September

Gooseberries: prune to create a goblet-shaped bush

powdery mould then you have mildew. Remove and burn all affected twigs.

Fruit is too small

Tree fruit and grape vines in particular need their crops to be thinned, everything comes to fruition at more or less the same time and the plant cannot produce sufficient resources to sustain all the fruit to maturation. Thinning is an essential activity if you want a good crop. Fruit can also be undersized if plants are suffering from lack of water. Irrigate the plant around its root system close to the soil – do not water the whole tree or bush.

No fruit

If you know the precise variety of tree or plant check to make sure that it does not need to be cross-pollinated and, if this is the problem, your only solution will be to plant a suitable pollinator nearby. If the tree/shrub was pruned a little too hard the previous year it may take a little while to recover and grow more fruiting wood. Alternatively it may be exhausted from years of fruiting. Prune in winter to stimulate new growth and feed it with a good mulch or top dressing every spring for several years.

Rotting fruit

If your fruit has rotten brown patches ringed with white fungus then you have brown rot. Remove and discard all damaged or withered fruit and dead shoots. Soft fruit are susceptible to grey mould which is caused by a fungus that attacks during flowering. By the time it is discovered on the fruit there is little you can do except to treat with chemicals during flowering the following year.

Holes in the leaves

These will generally be caused by caterpillars – hunt them down, shake the bush and pick them up when they fall to the ground and dispose of them. Sticky bands will make it difficult for them to climb back into the shrub.

Distorted leaves

This often indicates the presence of a virus, particularly in soft fruit where the virus is easily spread from plant to plant by greenfly and leaf hoppers. Remove diseased wood and burn. Attract natural predators in the garden so that the problem can be dealt with organically. Blackcurrants and gooseberries are prone to leaf spot. Pick off all affected leaves as well as all fallen leaves and burn.

Withered plants

This usually indicates the presence of disease; a lot depends on the plant in question. Raspberries suffer from cane blight and affected leaves and canes should be removed and burnt.

Cankers on wood

This indicates that the tree is diseased; apples are particularly susceptible to canker. A callous forms around the wound which will spread and encircle an entire branch. Die back can sometimes be seen on the shoot tips. Cut out and burn affected limbs. If the trunk is affected try to remove affected bark and paint the wound with a sealant.

Aphids

Plant companion plants such as nasturtiums, which aphids adore and chives which they avoid! Hoverflies and ladybirds will consume aphids given the chance, so attract them to the area with the poached egg plant (*Limnanthes douglasii*).

Caterpillars and beetles

Pick them off by hand, or shake them off the bush and grab them. Put sticky bands and grease bands around the trunks of trees or bushes to prevent them climbing back up.

Birds, butterflies and moths

The only solution to birds eating your crop is to net it. This does not involve constructing a permanent fruit cage – though they do make life very easy. For soft fruit you can just rig up some nets on bamboo canes and crawl underneath when you want to pick the fruit! Trees are harder and you may just have to get picking quickly, but you can try scare tactics – hang CDs, foil containers and lightweight plastic bags onto string and hang in the branches.

Snails and Slugs

Create barriers around vulnerable plants with sawdust, egg shells, sand or wood ash. Lay traps such as saucers of beer or upturned oranges to attract them, then dispose of them. Traps are available from garden centres.

Wasps

Wasps will want a share of your fruit crop – to distract them, half fill plastic bottles with fizzy pop or sugared water and hang up with string near the fruit. First make a small hole in the cap and screw it back on – or use foil to make a cover and make a hole in that. The wasps can get in, but not out and eventually they drown. Do remember though, they also dispose of a lot of insects so they are useful to have around.

ORGANISATIONS

GARDEN ORGANIC
Ryton Organic Gardens
Coventry
Warwickshire
CV8 3LG
www.gardenorganic.org.uk
A national charity researching principles and practice of organic gardening and a wealth of information. Garden Organic also operates a seed bank to ensure continuation of old and unusual plant varieties. Formerly Henry Doubleday Research Organisation.

GARDEN ORGANIC YALDING
Benover Road
Yalding
Nr. Maidstone
Kent
ME18 6EX
www.gardenorganic.org.uk
Linked to Garden Organic, Yalding shows gardens through the ages and has a beautiful organic fruit garden.

AUDLEY END ORGANIC KITCHEN GARDEN
Saffron Walden
Essex
CB11 4JG
A beautiful kitchen garden much as it would have been in late Victorian times.

BROGDALE HORTICULTURAL TRUST
Brogdale Road
Faversham
Kent
ME13 8XZ
www.brogdale.org
Home of the National Fruit Collection, stunning in spring when the trees are in bloom, if you need help deciding what apples, plums, pears, cherries, etc etc to grow this is the place to visit.

EAST OF ENGLAND APPLES AND ORCHARDS PROJECT
The School House
Rougham
King's Lynn
Norfolk
PE32 2SE
www.applesandorchards.co.uk
Masses of information on orchard fruit and a website which includes a where to buy directory, a local events calendar and training in cooking, regenerating old trees and all manner of desirable skills for enthusiasts.

GREAT DIXTER
Northiam
Rye
East Sussex
TN31 6PH
www.greatdixter.co.uk
Stunning gardens created by Christopher Lloyd and his family.

HEREFORD CIDER MUSEUM
21 Ryelands Street
Hereford
HR4 0LW
www.cidermuseum.co.uk
If you want to know more about cider making this is the place to visit.

ROYAL HORTICULTURAL SOCIETY
80 Vincent Square
London
SW1P 2PE
www.rhs.org.uk
The UK's leading gardening charity dedicated to advancing horticulture and promoting good gardening. A wealth of information online.

RHS GARDEN WISLEY
Woking
Surrey
GU23 6QB
www.rhs.org.uk
Fabulous RHS gardens and a great fruit garden.

The following website is also worth a visit:
www.freshfoodcentral.com
Lots of information about fruit and vegetables including a focus on what is in season

BIBLIOGRAPHY

PRINCIPLES OF HORTICULTURE by C.R. Adams, K.M. Bamford and M.P. Early, *published by Heinemann Newnes*

THE WARTIME KITCHEN AND GARDEN by Jennifer Davies, *published by BBC Books*

HARRY DODSON'S PRACTICAL KITCHEN GARDEN by Harry Dodson and Jennifer Davies, *published by BBC Books*

A GREENER LIFE, by Clarissa Dickson Wright and Johnny Scott, *published by Kyle CathieLtd*

THE PRICKOTTY BUSH by Montagu Don, *published by Macmillan*

BOB FLOWERDEW'S ORGANIC BIBLE by Bob Flowerdew, *published by Kyle Cathie*

THE FRUIT EXPERT by Dr D.G. Hessayon, *published by Transworld*

COMPLETE GUIDE OF HOME GARDENS by M. James, *published by Associated Newspapers Ltd*

ALLOTMENTS by R.P. Lister, *published by Silent Books*

GARDENER COOK by Christopher Lloyd, *published by Frances Lincoln*

PRUNING MADE EASY by M. Lombardi and C. Serra Zanetti, *published by Ward Lock*

FOOD FROM YOUR GARDEN *published by Reader's Digest*

ENCYCLOPAEDIA OF GARDEN PLANTS AND FLOWERS *published by Reader's Digest*

NEW ILLUSTRATED GUIDE TO GARDENING *published by Reader's Digest*

THE ENGLISH APPLE by Roseanne Sanders, *published by Phaidon in association with the Royal Horticultural Society*

DELIA'S KITCHEN GARDEN by Gay Search, recipes by Delia Smith, *published by BBC Books*

THE COMPLETE GARDENER by W.E Shewell-Cooper, *published by Collins*

PRACTICAL GARDENING AND FOOD PRODUCTION IN PICTURES by Richard Sudell, *published by Odhams Press Ltd*

GLOSSARY

BLEEDING If a tree is heavily pruned, or if pruned in spring when the sap is rising, it may lose a lot of sap and be weakened. Trees that need to be regenerated should be pruned over a period of three years, not all at once.

BUD The growing point of a plant from which leaves or flowers develop.

BUSH A type of fruit tree that has a short trunk, not to be confused with a soft fruit bush.

CENTRAL LEADER The trunk.

DORMANT Fruit trees and bushes are deemed to be dormant in late autumn and winter when growth is not occurring.

DROP The term applied to natural fruit fall, usually in June, by which method the plant reduces the size of its crop.

ESPALIER A form of fruit tree with horizontal branches trained in tiers and supported by wires.

FAN TRAINED A tree is trained to grow in a fan shape against a wall, a style favoured for peaches.

FORCING Growth stimulated ahead of the usual season.

FRUIT BUDS These are rounder and fatter than leaf buds.

FRUITING SPUR A branch that produces leaf shoots as well as fruit buds. Leaf shoots are pruned in summer to let light onto the fruit.

GRAFTING A method of propagation in which two plants are joined together to unite. The resulting plant takes particular characteristics of both plants.

GROWTH BUDS Growth or leaf buds are smaller and leaner than flower buds.

HEELING IN Temporary planting of a tree or plant.

LATERAL A side shoot on a leading shoot.

LEADING SHOOT Leaders are the current year's growth of wood. Leaders are pruned in winter.

LAYERING A method of propagation whereby a runner from the main plant is submerged in soil to encourage it to root.

MAIDEN A single stemmed tree in its first year after grafting.

MULCHING Placing a layer of compost or farm yard manure around specific plants to feed them and minimise water loss. A mulch can also refer to a layer of plastic or bark chips placed around plants to suppress weeds and minimise water loss.

PH BALANCE A scale by which the acidity and alkalinity of soil is measured.

POLLINATION The movement of pollen from one flower to another to effect fertilisation.

PROPAGATION A system of increasing plant stocks, including growing plants from seed, division, cuttings and runners.

PRUNING The process of controlling the growth of a healthy plant to improve shape, control growth, or to promote flowering and fruiting.

ROOTSTOCK Plants with strong and vigorous roots used for grafting fruit trees or ornamental blossom trees.

RUNNER A term used to describe a rooting stem thrown out by the parent, runners can be planted out and severed from the parent when they have rooted.

SELF FERTILE A tree or plant which produces flowers that can pollinate other flowers on the same tree.

SPUR A short lateral branch that produces many flowers and subsequently fruit.

STANDARD A tree with a minimum trunk height of 1.8 metres.

SUCKER A shoot coming from the base of a plant or from the roots below.

SUB-LATERAL A side shoot on a lateral.

TENDER A plant that is not hardy and which cannot withstand frost.

TILTH The crumb-like structure of soil. Sewing seeds requires a fine tilth – i.e. soil that has been well prepared and raked.

TIP BEARING The term applied to trees that carry some of their crop on the ends of their shoots.

INDEX

acidity & alkalinity,
    soils 24
alcohol 11–12, 65–6
annual tasks 33
aphids 22, 88
apples 9, 11, 12, 16,
    49, 55–7
diseases 84, 87
preserving 43, 44, 45,
    46, 48
pruning & forms 18,
    19, 35, 39, 40, 49,
    56
apricots 11, 12, 19,
    39, 43, 44, 58–9

barrier techniques,
    pest control 42,
    81–2, 88–9
big bud 81
birds 13, 42, 81, 88
blackberries 16, 20,
    40, 44, 45, 71–2, 85
blackcurrants 11, 16,
    20, 22, 67–9, 81,
    85, 87
bottling 11, 43, 47–8
brewing 11–12
brown rot 86
bushes 15, 18, 20, 30,
    40

cane blight 87
canker 87–8
caterpillars 22, 42,
    81, 84, 87
chalk soils 24
cherries 11, 16, 39,
    40, 45, 59–61

pruning & forms 18,
    19, 39, 40, 49, 60
children 11, 12, 15,
    67
chutneys 11, 46–7
clay soils 23, 27
commercially–grown
    fruit 8–10, 21, 53,
    69, 84
companion planting
    14, 56, 82, 83, 88
compost 22, 25–7, 34
cordons 15, 18–19,
    20, 29, 41
cuts, pruning 37–8

damsons 11, 40, 44,
    46, 63
diseases 81, 84, 86–8
drainage 27–8
drying 48

elderberries 74
equipment & tools
    37, 78–80
espaliers 15, 19, 20,
    29, 41

fans 15, 19, 29, 60,
    62
farmyard manure 25,
    27, 34
feeding & fertilisers
    24–5, 34
figs 11, 19, 22, 34,
    49, 52
forcing rhubarb 71
forked cordons 15
freezing 11, 43

frost pockets 28
fruit, technical
    definition 10
fruiting problems
    86–7

gages 61–2
gooseberries 11, 16,
    22, 46, 69–70
pests & diseases 82,
    84, 87
pruning & forms 18,
    20, 40, 49, 70, 85
grapes 11, 12, 22, 41,
    50, 75–7
green composts 26–7
grey mould 86–7

half standards 17, 20,
    30
hard pruning 35, 37,
    39
heeling in 31
horsetail fungicide
    82–3

insects 13, 14, 22, 42,
    81–4, 87, 88, 89
irrigation 28, 86

jams & jellies 11,
    42–6

leaf problems 84, 87
light pruning 35, 37
loam soils 23
loganberries 16, 20,
    40

manure 25, 27, 34
mildew 84
Morello cherries 50, 60, 61
mould 86–7
mulberries 12, 57–8
mulching 27, 34

nectarines 8, 19, 63–5
nitrogen 16, 24, 34

old-fashioned varieties 35–6, 49
organic gardening 21–2

paths 15
peaches 11, 18, 40, 45, 49, 63–5
pears 8, 11, 12, 45, 66–7, 82, 84
pruning & forms 18, 19, 39, 40, 67
pectin 44–5
pests 13, 14, 22, 42, 81–4, 87, 88–9
pH balance 24
phosphorus 24, 34
plant selection 16–17, 35–6, 49–50
planting 19–20, 30–2
plums 8, 11, 12, 16, 44, 46, 61–2
pruning & forms 18, 19, 37, 40, 49, 62
pollinators 86
posts & wires 29
potash/potassium 16, 24, 25, 34
predators 13, 14, 22, 42, 82–3
preserving fruit 11, 17, 42–8, 65–6

pruning 15, 16, 34–41, 85, 86
see also specific forms (eg espaliers); specific fruits (eg peaches)
purchased fruit 8–10, 21, 53, 69, 84
pyramids 15
pyrethrum insecticide 82

quince 51

raspberries 11, 16, 22, 50, 72–4, 82, 87
preserving 44, 45
pruning 20, 29, 40, 73
redcurrants 11, 16, 67–9
preserving 44, 45
pruning & forms 18, 20, 40, 49, 68
rhubarb 44, 70–1
rootstocks 18, 50
rot 86
rusts 84

sandy soils 23
seaweed 25
silt soils 23–4
silver leaf 62
situation, fruit garden 11, 14–15, 21, 28, 50
sloes 11, 65–6
slugs 42, 88–9
soft fruit, definition 49, 50
soil 21, 22, 23–4, 27–8

spacing requirements 11, 14, 16
spur-bearing fruit, pruning 35, 56, 67
staking 30
standards 17, 20, 30, 49
sterilising bottled fruit 47–8
strawberries 11, 12, 27, 30, 50, 53–5
preserving 43, 44, 45
purchased 7, 8, 53
summer pruning 39, 40, 41
supports 29–30
symptoms, pest & disease problems 84, 86–9

taste 7, 10, 11, 69
tayberries 49
thinning 39, 56, 86
tip-bearing fruit, pruning 35, 56, 67
tools & equipment 37, 78–80

varieties, selecting 16–17, 35–6, 49–50

wasps 89
watering 28, 86
weed control 23, 27
wineberries 49
winter pruning 38–40, 86
wire supports 29
withering 87
wood ash 25

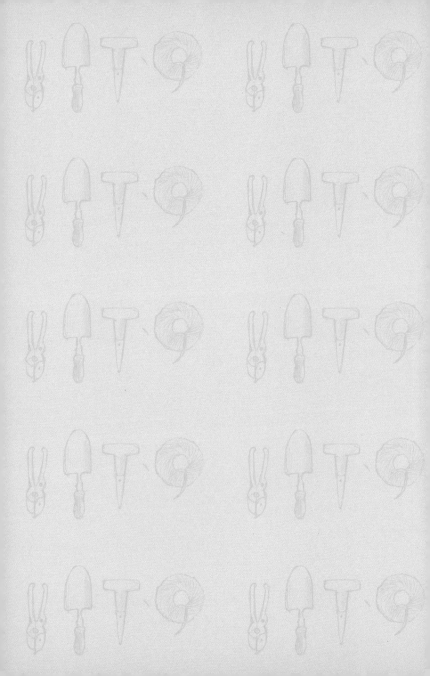